Springer Tracts in Mechanical Engineering

For further volumes:
http://www.springer.com/series/11693

Paolo Castaldo

Integrated Seismic Design of Structure and Control Systems

Springer

Paolo Castaldo
Civil Engineering
University of Salerno
Salerno
Italy

ISSN 2195-9862 ISSN 2195-9870 (electronic)
ISBN 978-3-319-02614-5 ISBN 978-3-319-02615-2 (eBook)
DOI 10.1007/978-3-319-02615-2
Springer Cham Heidelberg New York Dordrecht London

Library of Congress Control Number: 2013950671

Printed on acid-free paper

Springer is part of Springer Science+Business Media (www.springer.com)

Preface

Passive structural control techniques are generally used as seismic rehabilitation and retrofit methodologies of existing structures. Most of the advanced research and applications of structural control in civil engineering has been focused on the analysis of existing structures integrated with passive, either hybrid or active, control devices in order to improve the structural performance under strong earthquakes. In all cases, both the structure and control system are therefore designed separately and only subsequently integrated by following the principles of a performance-based design. An example is the case of "Millennium Bridge" whose structural functionality has only been restored through the use of viscous dampers to reduce the resonance phenomena.

An exciting consequence of structural control research is that it also opens the door to new possibilities in structural forms and configurations, such as slender buildings or bridges with longer spans without compromising the structural performance. This can only be achieved through the integrated design of structures with control elements as an integral part. In recent years, integrated optimal structural/control system design has been acknowledged as an advanced design methodology for space structures, however, not many studies and applications can be found in civil engineering.

In this work, with specific reference to the supplemental passive energy dissipation through viscous or viscoelastic devices, the possibility of achieving seismic protection through the integration of elastic resources of a framed structural system as well as viscoelastic ones of a dissipative bracing system has been investigated. The innovative aspect, therefore, consists of considering the viscoelastic damping resources as design variables to control the dynamic response.

A procedure for the integrated design of a framed structural system equipped with viscoelastic/viscous damper-brace component is therefore proposed and developed, in order to achieve an expected seismic design performance, by following the basic principles of the displacement-based seismic design and explicitly considering the dynamic behavior both of the structural system as well as the dissipative system. The choice of the optimal design is made by determining the combination of the design variables, which minimizes a cost index that is evaluated considering the relative cost between the elastic and viscoelastic dissipative resources.

The structural optimization, developed in this work, made it possible to obtain new optimized solutions of the design problem for fixed shape and structural topologies through the integrated use of dissipative resources produced by dampers, resulting in slender structural systems with a high seismic performance.

This methodology is, finally, a response to the futuristic idea to ensure an adequate seismic performance for structural solutions including innovative materials, charactrerized by a high slenderness.

Acknowledgments

I would like to express my gratitude to all those who have supported and encouraged me with their advice, wisdom, and affection allowing me to grow on a scientific, professional, as well as human level.

In primiis, I thank Prof. Bruno Palazzo, who has given me the opportunity to investigate numerous scientific and practical issues in structural and seismic engineering. With his teachings and advice, he has always been a guide for me.

A heartfelt thanks to my friend, Eng. Massimiliano De Iuliis, who, with his scientific teachings and moral support, has always been a unique and special person. So many were the times spent with him and the scientific comparisons.

A heartfelt and sincere thanks to my friend, Eng. Michele Calvello, for his advice and teachings as well as my friends and academic colleagues. I thank Prof. Anna Amirante who, during High School, allowed me to further my scientific education.

I would also like to thank all the true and sincere people who I have met over the years. In particular, I thank Marianela for all the days, full of smiles, joys, and emotions lived together.

With great joy, I thank my cousin Stefano who has supported me in every circumstance of my life and study.

A big and special thanks to my twin brother Francesco, with whom I have always shared every aspect of my life and who has always supported me in all my choices.

Many thanks to my parents "Antonio and Maria" who allowed me to obtain the results achieved, always giving me the right advice and their support, together with all my cousins and relatives, in particular, Raffaele and Giuseppina who have considered me and my brother as their two other sons. With affection and respect, I also thank my uncles Mario Palmieri and Mario Mello for their valuable suggestions.

Finally, a warm thanks to my grandparents, for their confidence and consideration shown toward me.

Paolo Castaldo

Contents

Chapter 1
Passive Structural Control

Abstract In this chapter, structural control techniques aimed at reducing the vulnerability of existing or new structures, built in medium and high seismic risk areas, are described. The description is focused on the passive control and, in particular, on the passive supplemental energy dissipation techniques. The effectiveness of the largest and most popular dissipation devices, such as hysteretic and viscous, by highlighting the significant reduction of seismic demand on the main structure to be protected through the evaluation of earthquake damage indices is briefly described.

1.1 Structural Control Techniques

After several destructive earthquakes in recent years in Italy (Friuli 1976, Campania and Basilicata 1980, Marche and Umbria 1997, L'Aquila 2009 and Emilia 2012), the need to redefine the design strategies and codes for the structures sited in high seismic risk areas has been highlighted. Deaths from earthquakes are almost always associated, as in the case of landslides, to damage suffered by buildings, dams, bridges and other human settlements. Unfortunately, the seismic events are characterized by a frequency that is difficult to predict and quantify in space and time due to the randomness of various seismological factors. An average of about 200 high magnitude events occur every decade, of which 10–20 % in the oceanic crust without causing problems to human settlements. Others occur in areas far from the cities and settlements and likewise cause very few problems. The biggest problems arise when an earthquake strikes densely populated areas. It is useful, at this point, to introduce the concept of risk. The risk is the threat to any property or the area's activities following the occurrence of a phenomenon of assigned intensity or return period T_R. It can be expressed as the product of three terms:

$$R = P \cdot V \cdot E \tag{1.1}$$

where P is the probability of occurrence of a phenomenon characterized by a return period (the return period is the average number of years that elapses

P. Castaldo, *Integrated Seismic Design of Structure and Control Systems*,
Springer Tracts in Mechanical Engineering, DOI: 10.1007/978-3-319-02615-2_1,

between two successive events of equal intensity), V is the vulnerability of the exposed properties and depends mainly on the type of infrastructure and services in the area; finally, E (exposure) represents the value of the goods exposed. The risk can be reduced in general by acting on one of the three term. However, in the case of seismic events, it is evident that it is not possible to act on the frequency of occurrence, i.e. on the dangerousness. Vulnerability is the only factor that the codes, design criteria and design strategies can reduce. Structural damage observed during several seismic events clearly demonstrates that the structural design determines the response of a building to the event. A structure characterized by a global symmetry and regularity both in elevation and in plan presents a more regular distribution of the seismic actions unlike the irregular systems in which the dynamic response increases in correspondence with the perimetral elements. Other problems may occur such as in tall structures since it is possible that the seismic action generates considerable overturning moments, i.e. compression and tension considerably increases in the vertical load-bearing elements of the edge, and in the structures with remarkable development in plant whose seismic response can be influenced by the topographic and stratigraphic variability with the consequence of not having a synchronic input in the different parts of the structure. Furthermore, due to the motion of the soil, a further problem, that could be significant, is the torsion caused by the eccentricity of the spatial configuration of the structure due to a non-coincidence between the center of mass and that of stiffness.

Due to the high level of structural damage observed in the post-earthquake phases, the belief that it is not possible to neglect forecasting procedures and capitalization of the costs of maintenance, restoration and consolidation work, which are necessarily considered during disastrous events as a function of the vulnerability of structures has been confirmed. An earthquake is not only a risk to lives, but can also involve severe damage to the architectural heritage of the area affected, with any form of retrofitting not always being easy in terms both of the costs to be incurred as well as the time. The experiences of other countries, such as Japan, have shown that the side effects mentioned above can be easily avoided as long as appropriate design techniques are used, with an extra initial cost that is extremely negligible in relation to the cost of any extraordinary maintenance and consolidation due to damage caused by earthquakes with high return period.

Thus, according to traditional seismic protection, buildings located in high seismic risk areas must have the following requirements:

a. the structural elements of the main system must have strength characteristics so as to overcome without damage medium intensity seismic events, which can potentially affect the construction for at least once during its useful life;
b. all the (primary and secondary) members of the structure must have ductility to dissipate without collapse mechanisms, the input energy of the high intensity earthquake events.

This ensures the safety of the occupants in the case of events characterised by a lower probability of occurrence and, to be overcome without plastic deformation, which would require an uneconomical size of the structure. In this case, damage to both the bearing and the complementary elements is acceptable, due to the dissipative mechanisms.

These strength and ductility benefits can be reached in modern structures, even without the aid of anti-seismic devices, provided that some basic design and execution rules are observed.

On the other hand, buildings designed according to the old codes, such as masonry buildings and monuments, do not have a high ductility and, in many cases, even traditional strengthening methods do not allow to significantly improve the ductility due to being aimed primarily at obtaining an increase of resistance in the elastic range. Moreover, it should be considered that the level of consolidation usually adopted in these cases does not allow to reach a rigorous seismic retrofit, due to both the high costs necessary as well as the significant perturbation produced by some technological consolidation systems (for example, the injections of cement mortar or resin), which should be avoided when rehabilitating historical buildings and monuments.

In order to overcome these problems and achieve a satisfactory seismic response, or to reduce the vulnerability of a structure, it is possible to dissipate energy through the plastic hinges or adopt structural control techniques. The methods and techniques used to control the structural behavior under dynamic loads due to wind or earthquake have had a significant development in the last decade (Soong and Dargush 1997; Symans et al. 2008). In general, structural control can be divided into three classes: active, semi-active and passive control. The structural control techniques, that are called active control, hybrid and semi-active, can be considered a natural evolution of passive control technologies (Soong and Spencer 2002).

The possible use of active control systems and some combinations of passive and active systems as a means of structural protection against seismic loads has received considerable attention in recent years. The active control systems, hybrid and semi-active devices are force delivery devices integrated with real time processing controllers and sensors within the structure. The control of the forces in an active control system is typically generated by electromechanical or electrohydraulic actuators based on either feedback information of the measured response or the feedforward information of external excitation (Symans and Constantinou 1997). The recorded measurements of the response and/or excitation are monitored in a controller that, based on a predetermined control algorithm, determines the appropriate control signal directed to the actuators. The generation of the control forces through the electrohydraulic actuators requires considerable power resources, which are of the order of tens of kilowatts for small structures and may reach many megawatts for tall structures. The first effect of some experimentally tested control systems was to change the level of damping with a minor modification of the stiffness. An overview of the active structural control is contained in several states of the art in current scientific literature (Soong and Constantinou 1994; Fujino et al. 1996).

Finally, a control system can be defined semi-active as a system that typically requires a small external energy resource for operations. It uses the motion of the structure to develop control forces, the intensity of which can be controlled from the external power source (i.e. magneto rheological or electro rheological) (Symans and Constantinou 1999; De Iuliis and Castaldo 2012). The control forces are based on feedbacks that come from sensors which measure the excitation and/ or the response of the structure. The feedback from the structural response can be measured in a location that is remote from one of the control of the semi active system.

Passive control, in turn, can be divided into four categories: tendon control, base isolation, passive aerodynamic control, passive damper control. The latter category includes viscous fluid, viscoelastic, frictional and hysteretic devices.

Among the different categories, it is worth mentioning base isolation (Kelly 1997), a technique that involves inserting an isolation level between the superstructure and foundation to obtain a more convenient variation of the characteristics of the dynamic response than a "fixed base" structure. The insulators are classified into elastomeric rolling and sliding.

To significantly reduce the stress states in the structure, it is necessary to minimize the inertia forces due to the earthquake. Since these depend on the structural mass and the acceleration induced by the absolute ground, the scope can be reached by reducing either the mass or acceleration on the structure. The base isolation brings down the acceleration on the system through a modification of the dynamic response of the system. The decoupling of motion, that follows from the interposition elements with high horizontal deformability and high vertical stiffness between foundations and the superstructure, usually involves the filtering of frequencies with a higher energy content, allowing the ones, characterized by lower values and usually with a lower energy content, to pass on. In this way, the first vibration mode of the structure is configured (Fig. 1.1) with a deformed shape

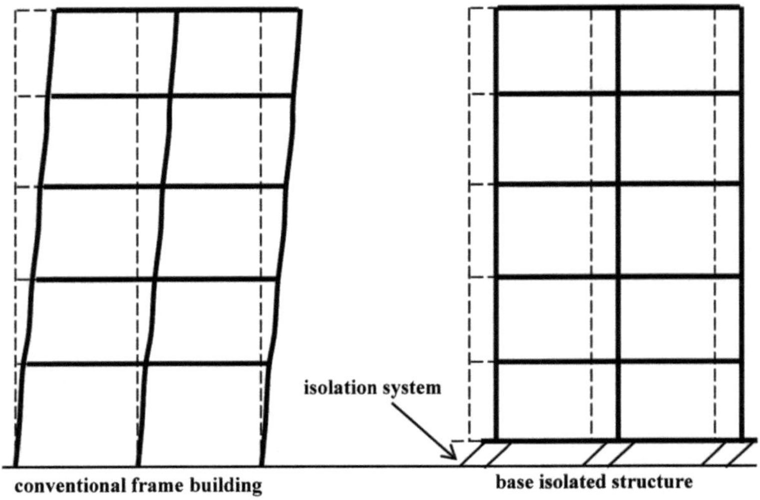

Fig. 1.1 Deformed shapes under an earthquake action of both a conventional frame building and a base isolated structure

similar to that of a rigid body that moves slowly (translation) above deformable devices, with very low absolute accelerations and very modest interstory deformations (no damage). Thus, it is associated to an excited mass that is generally higher than 90 %, with it identifying the response of the structure during an earthquake. The structural behavior is similar to that of a simple degree of freedom (SDOF) system, with the mass coincident being the total mass and the stiffness of the system that coincides with that of the isolation system. The participation of higher modes is greatly reduced in the evaluation of both the internal deformation and the accelerations. If, due to the almost total absence of interstory deformations, this technology practically avoids damage to non-structural elements (cladding, partitions, fixtures), as a result of the filtering of higher frequencies, it also makes it possible to avoid that the "content" suffers heavy acceleration and thereby contributes to its effective protection. It follows that everything depends on the reliability and proper functioning of the isolation level to be certified with the appropriate quality and designed through detailed linear or non-linear analysis depending on the type of isolators chosen.

Another remarkable technique that can be used to obtain an efficient seismic response is the installation of energy dampers. The purpose of the passive control through the use of passive (supplemental) energy dissipation devices (PED, Passive Energy Dissipation) is to reduce as much as possible the damage in the structural system by dissipating the greatest possible amount of energy input (Soong and Dargush 1997; Symans et al. 2008; Soong and Constantinou 1994; ATC 1994; Constantinou et al. 1996) through the use of specific devices, positioned in parallel to the main structure or in a seismic isolation system.

Depending on their conformations, these systems can also increase the stiffness and strength to the structure upon which they are installed. A passive control system does not require an external power source to be operational, rather, the motion of the structure is used to produce a relative motion between the ends to which the dissipation devices are connected. The passive supplemental energy dissipation systems can have various shapes and dissipate energy through various mechanisms including the yield strength of the steel, the viscoelasticity of materials such as rubber, the compression of a viscous fluid forced to pass through the holes and sliding friction. It follows that every type of passive supplemental energy dissipation devices is based on the specific physical–mechanical properties that influence, in a different way, the dynamic response of the main structural system. They absorb the earthquake energy, reducing the effects on the elements of the main structure. After the event, these devices, which do not have a static function, are replaced, leaving the structure undamaged. The main difference between the base isolation and dissipation systems is easily understood by analyzing how they affect the seismic demand in the structure in which they are installed (Fig. 1.2).

The base isolation owes its effectiveness to the fact that the design spectrum has spectral peaks maximum for low periods, between 0.2 and 0.6 s, while the isolation system increases the period of structure, bringing it out of this peak range. The dissipation system moves the period of the structure towards lower values,

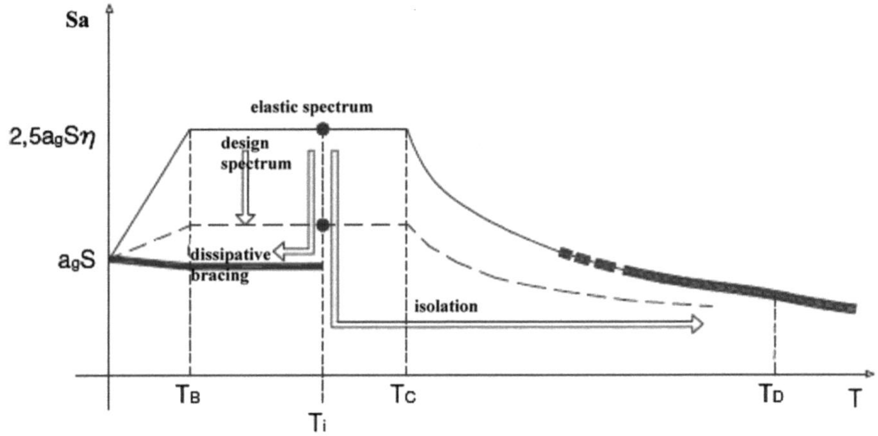

Fig. 1.2 Effects due to passive protection tecnhiques in Sa-T plan

thus increasing the seismic action on the structure, but which, through the dissipative mechanisms that are developed, is able to strongly reduce the drifts and, consequently, any structural damage. In Table 1.1 the three classes for seismic protection are shown. The three major classes of control systems are sometimes combined to form so-called hybrid control systems. Figure 1.3a–e show the process schemes for the three types of structural control.

1.2 Passive Energy Dissipation Systems

Passive energy dissipation systems for seismic applications have been under theoretical development since the mid-1980s with a rapid increase in implementations starting in the mid-1990s. A large number of passive control devices have been installed in structures to improve their performances under seismic actions. Specifically, in North America, these devices have been implemented in numerous buildings and bridges, both for the adaptation of existing and new buildings (Soong and Spencer 2002). Figure 1.4 shows the distribution of these buildings as a function of the year in which PED systems were installed.

Table 1.1 Structural protective systems

Seismic isolation	Passive energy dissipation	Active control
Elastomeric bearings	Metallic dampers	Active bracing systems
Lead rubber bearigns	Friction dampers	Active mass dampers
Sliding friction pendulum	Viscoelastic dampers	Variable stiffness or damping systems
	Viscous dampers	
	Tuned mass dampers	
	Tuned liquid dampers	

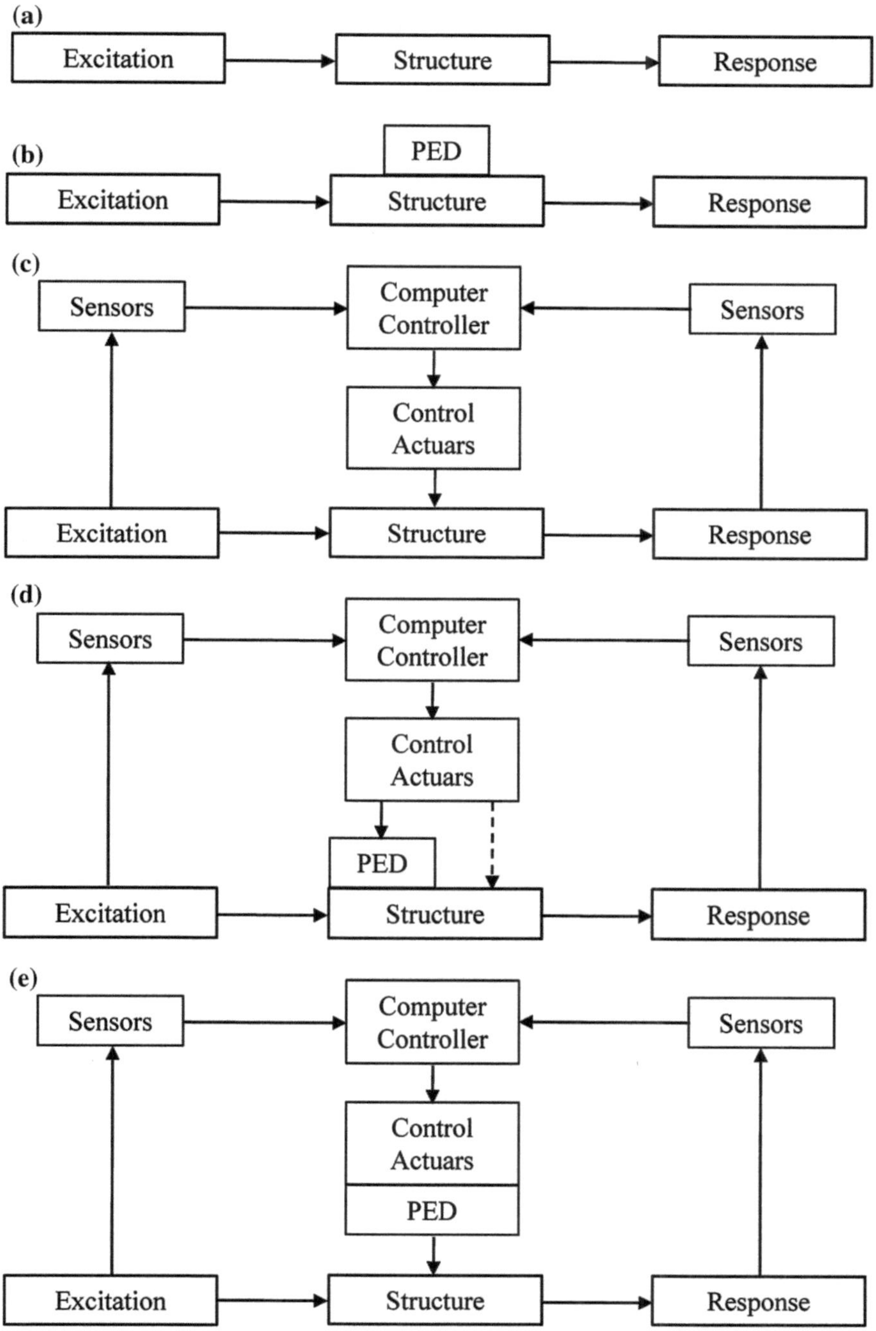

Fig. 1.3 **a** conventional structure; **b** structure with passive energy dissipation (PED); **c** structure with active control; **d** structure with hybrid control; **e** structure with semi-active control (modified from Soong and Spencer 2002)

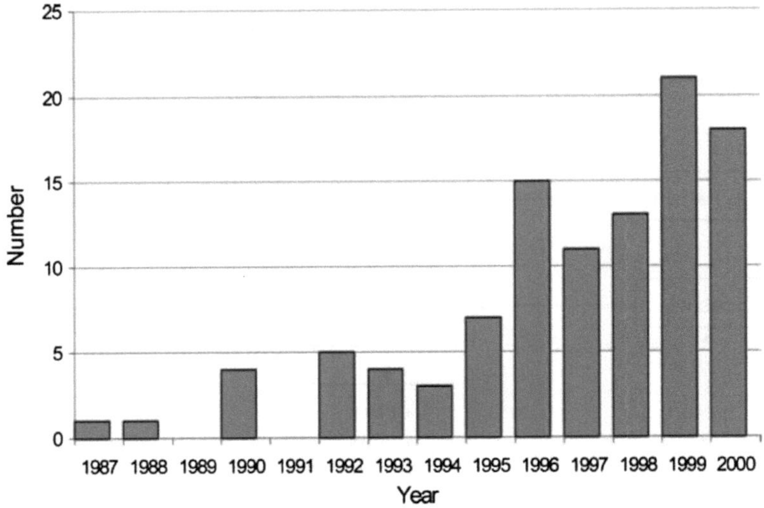

Fig. 1.4 Implementatios of passive energy dissipation in North America for seismic applications (modified from Soong and Spencer 2002)

The main function of a passive energy dissipation system is to reduce the inelastic energy dissipation demand on the framing system of a structure (Constantinou and Symans 1993; Whittaker et al. 1993) and subsequently damage on the main system. A large number of passive dissipation devices are either commercially available or under development. Devices commonly used for the seismic protection of structures are viscous fluid dampers, viscoelastic (solid) dampers, friction and metallic dampers. Other devices that could be classified as passive energy dissipation devices include tuned mass and liquid dampers, both of which are mainly used to control wind vibration.

As already stated, the main reason for using passive energy dissipation devices in a structure is to limit damaging deformations in structural components. The degree to which a certain device is able to achieve this depends on the inherent properties of the basic structure, the properties of the device and its connecting elements, the characteristics of the ground motion as well as the limit state being investigated. Given the large variations in each of these parameters, it is usually necessary to carry out an extensive series of non-linear response-history analyses to determine which particular passive energy dissipation system is best suited for a given case.

Consider the lateral motion of the basic single-degree-of-freedom (SDOF) model (Soong and Dargush 1999) (Fig. 1.5), consisting of a mass m, supported by springs with total linear elastic stiffness k, and a damper with linear viscosity c.

This SDOF system is then subjected to a forced vibration, $f(t)$. The excited model responds with a lateral displacement $x(t)$ relative to the ground $x(t)$, according to the following relation:

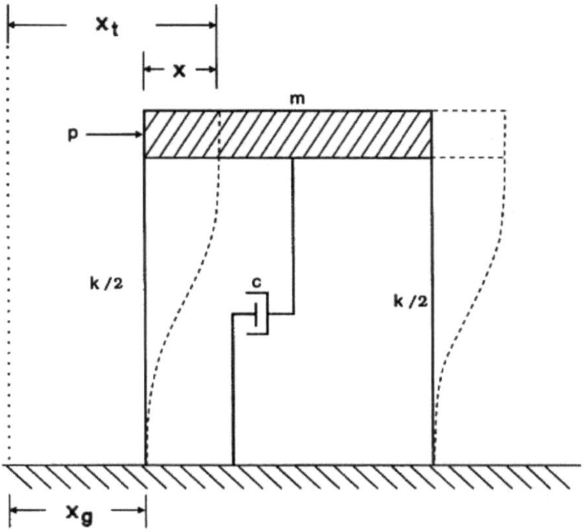

Fig. 1.5 SDOF system (modified from Soong and Dargush 1997)

$$m\ddot{x} + c\dot{x} + kx = f(t) \quad (1.2)$$

where *f(t)* can be represent, for example, a wind or seismic load. In the case of seismic load, it is representative of the inertial force, proportional to the mass *m*, caused by the ground acceleration, i.e.:

$$f(t) = -m\ddot{x}_g(t) \quad (1.3)$$

and $\ddot{x}_g(t)$ is ground acceleration.

Consider, then, the addition of a generic passive damper element into the SDOF model, as indicated in Fig. 1.6.

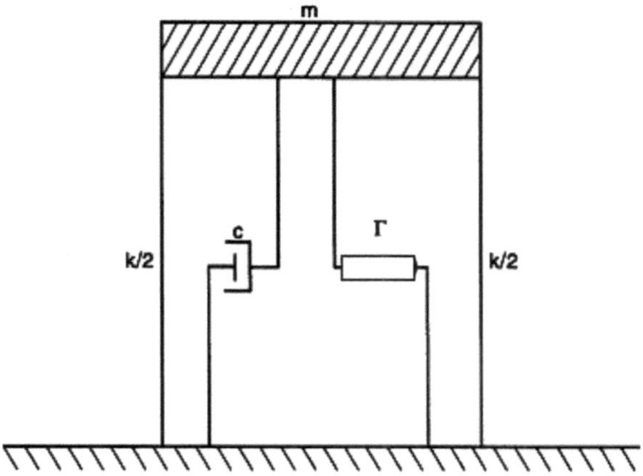

Fig. 1.6 SDOF system with passive energy dissipation (modified from Soong and Dargush 1997)

The response of the system is now influenced by this additional element, which must be characterized in terms of a suitable macroscopic force–displacement model. The symbol Γ in Fig. 1.6 represents a generic integrodifferential operator, so that the force corresponding to the passive device can be written simply as Γx (Soong and Dargush 1997). This permits quite general response characteristics, including displacement, velocity, or acceleration-dependent contributions, as well as hereditary effects. The motion equation for the damped SDOF model then becomes:

$$m\ddot{x} + c\dot{x} + kx + \Gamma x = -(m + \bar{m})\ddot{x}_g \tag{1.4}$$

with $\bar{m}$ representing the mass of the passive element.

The specific form of the term Γx should be indicated so that the equation can be solved and it is, obviously, dependent on the type of device. A typical force–displacement relationship for rate-independent elastic-perfectly plastic elements is depicted in Fig. 1.7, defined in terms of an initial stiffness $\bar{k}$ and a yield force $\bar{f}_y$.

Assume for illustrative purposes that the base structure has a viscous damping ratio $\xi = 0.05$ and that a simple massless yielding device is added to serve as the passive element, characterised by the above-mentioned force–displacement relationship for the term Γ. Considering the passively damped SDOF model subjected to harmonic loading and examined under steady-state conditions after all the transients have dissipated, Fig. 1.8 presents the amplitude of the displacement response versus forcing frequency. In order to clearly identify the effect of the added passive damper, the normalization is accomplished in terms of the natural frequency $\omega_0 = (k/m)^{1/2}$ and static response $x_{st} = p_0/k$ of the base structure (Soong and Dargush 1997). In this figure, the uppermost curve represents the response of the SDOF system with $\xi = 0.05$ and without devices. The other two curves illustrate the effect of adding a yielding device to the system. It is evident that the peak response, near resonance, is greatly reduced due to the energy dissipation devices.

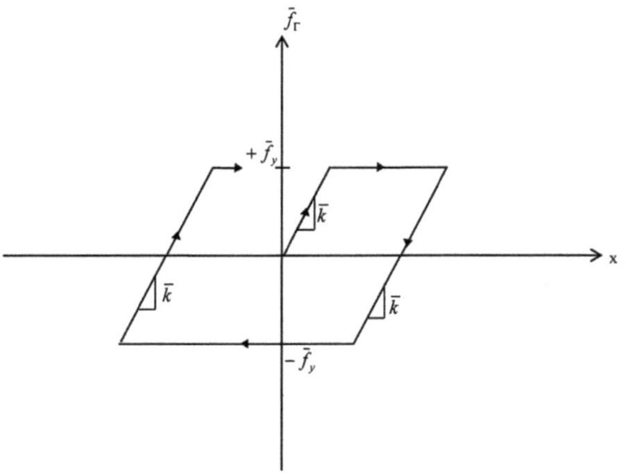

Fig. 1.7 Force-displacement model for elastic-perfectly plastic passive element (modified from Soong and Dargush 1997)

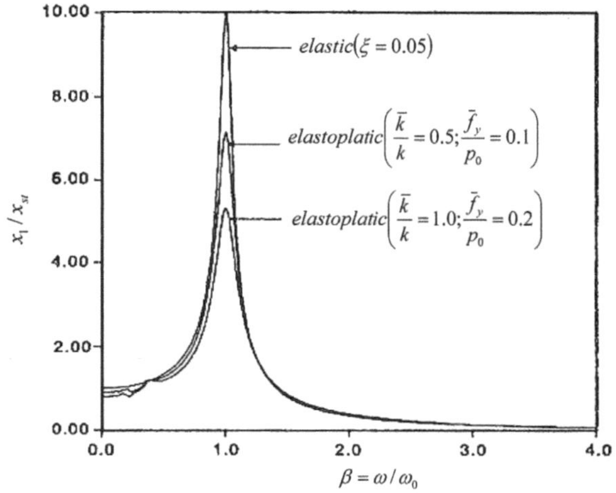

Fig. 1.8 Amplification factors for harmonic excitation with passive element (modified from Soong and Dargush 1997)

Then, the passively damped SDOF structure has been analyzed for response due to the 1940 El Centro S00E ground motion (Soong and Dargush 1997). The initial stiffness of the elastoplastic passive device is specified $\bar{k} = k$, while the yield force, $\bar{f}_y$, is equal to 20 % of the maximum applied ground force:

$$\bar{f}_y = 0{,}20 \max\{m|\ddot{x}_g|\} \tag{1.5}$$

The resulting relative displacement and total acceleration time histories are presented in Fig. 1.9a, b relating to the SDOF system and in Fig. 1.10a, b to the passively damped SDOF system.

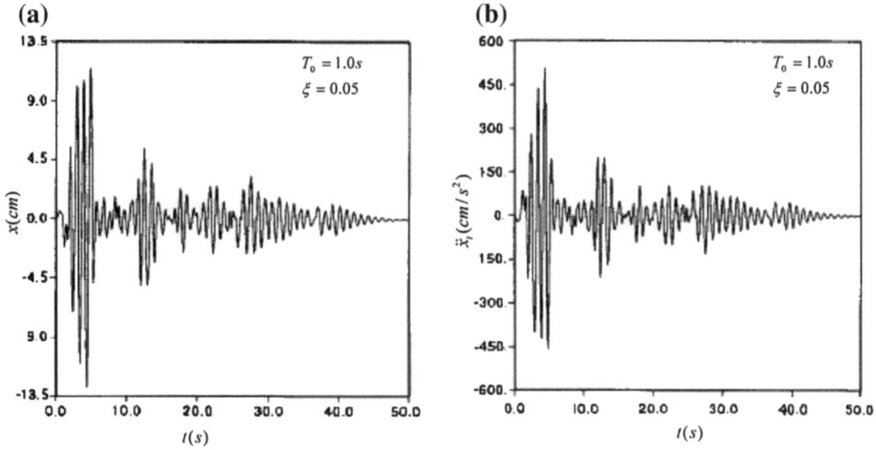

Fig. 1.9 Force-displacement response for El Centro 1940 excitation with passive element: **a** displacemnets; **b** accelerations (modified from Soong and Dargush 1997)

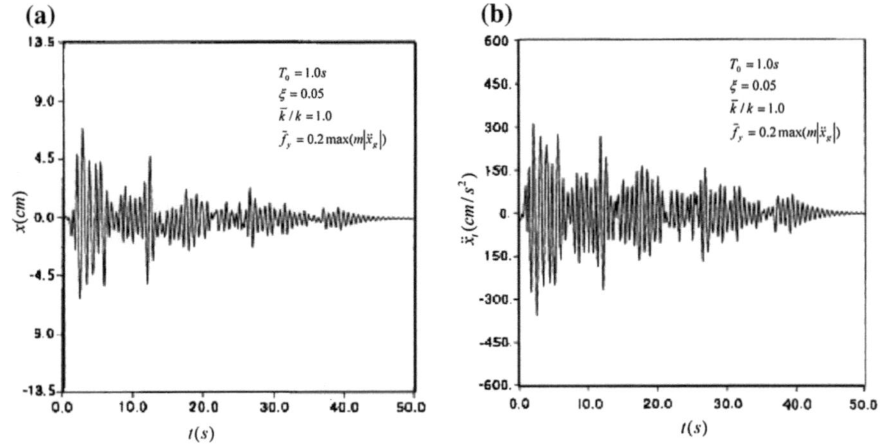

Fig. 1.10 1940 El Centro time history response for SDOF system with passive elements: **a** displacements; **b** accelerations (modified from Soong and Dargush 1997)

Again, there is a significant reduction in response compared to that of the base structure without the passive element. A similar reduction can be achieved using viscous devices (Soong and Dargush 1997), in Fig. 1.11a, b, the typical force–displacement cycles for the viscous and hysteretic dampers are shown, respectively.

In this case, the size of these loops indicates that a significant portion of the energy is dissipated in the passive device. This tends to reduce the forces and displacements in the primary structural elements, which of course is the purpose of adding the passive device.

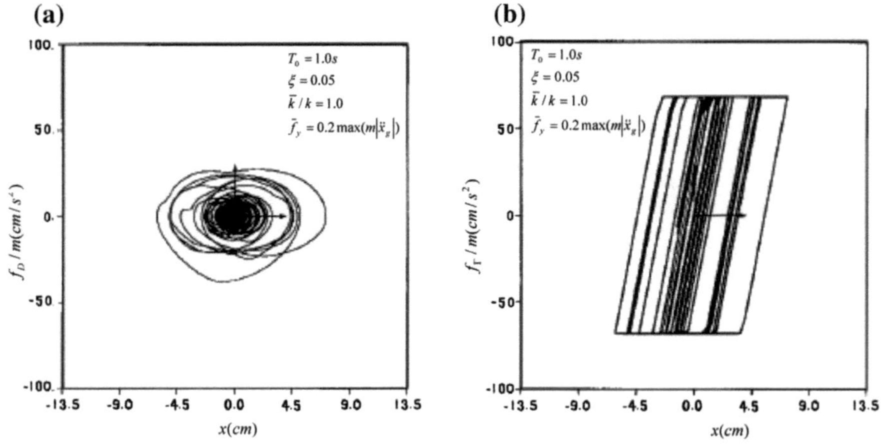

Fig. 1.11 1940 El Centro force–displacement response for SDOF system with passive element: **a** viscous damper; **b** passive damper (modified from Soong and Dargush 1997)

An energy balance can be obtained by integrating every term in the motion Eq. (1.4) on the entire history of relative displacements:

$$E_K + E_D + E_S + E_H + E_P = E_I \tag{1.6}$$

where:

$$E_K = \int m\ddot{x}\,dx = \frac{m\dot{x}^2}{2} \tag{1.7}$$

$$E_D = \int c\dot{x}dx = \int c\dot{x}^2 dt \tag{1.8}$$

$$E_S = \int kxdx = \frac{kx^2}{2} \tag{1.9}$$

$$E_I = -\int m\ddot{x}_g dx \tag{1.10}$$

$$E_P = \int \Gamma x dx \tag{1.11}$$

E_K is the instantaneous kinetic "relative" energy of the moving mass; E_D is the cumulative viscous damping energy due to the damping of the basic structure; E_S is the instantaneous strain energy stored by the structure; E_H is the cumulative hysteretic energy; E_P is the cumulative viscous damping energy due to the damping of dissipation devices. The sum of these energies must balance the cumulative input energy E_I of the seismic event. It is noted that each term is actually a function of time, and that the energy balance is obtained at each instant during the entire duration of the motion. At the end of the earthquake ($t = t_f$), the kinetic energy is zero, the (elastic) strain energy is zero for an elastic system (it is nothing or almost nothing for an inelastic system), and the cumulative hysteretic energy is equal to the energy demand $E_H(t_f) = E_{Demand}$.

A structure designed according to the current international codes has a non-linear response during a seismic event of considerable intensity. It follows that the engineering design must be based on the ductility of structural elements to prevent collapse, accepting the fact that the damage can take place. If the seismic input energy, in fact, exceeds the capacity of the structure to store and dissipate energy through the mechanism of elastic deformation, some parts of the structure suffer damage linked to plastic deformations. This can be prevented by using passive dissipation devices.

As a significant example of the effects of a passive device on the energy response of a structure, the results of some tests of a reinforced concrete framed construction of three floors in scale 1:3 conducted by Lobo et al. (1993) are illustrated.

Figure 1.12a, b show the measured response of the structure caused by the 1952 Taft N21E ground motion with PGA equal to 0.2 g.

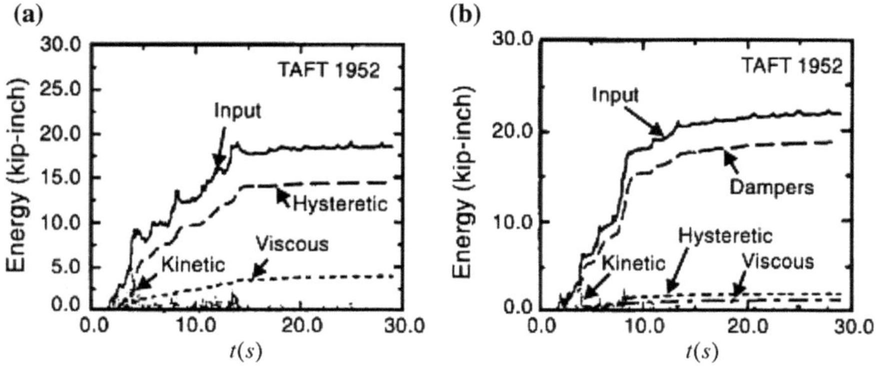

Fig. 1.12 Energy response of test structure **a** without passive devices; **b** with passive devices (modified from Soong and Dargush 1997)

A considerable portion of the input energy is dissipated through hysteretic mechanisms, which damage the primary structure through the cracking and formation of plastic hinges. It is noted that such damage is minimized by the addition of a set of viscoelastic devices. In fact, in spite of the input energy being slightly increased, the devices dissipate a significant portion of the total energy, thereby protecting the primary structure.

Another exhaustive example of a structure with a passive dissipation devices has been proposed by Symans et al. (2008), considering the motion of the base of the 1940 S00E El Centro earthquake (Fig. 1.13). This example is discussed below to clearly illustrate the basic principles of energy dissipation systems for seismic applications.

The idealized structure consists of a one-story, one-bay moment resisting frame having weight W_0, mass M_0, lateral stiffness K_0, and lateral strength Y_0. The

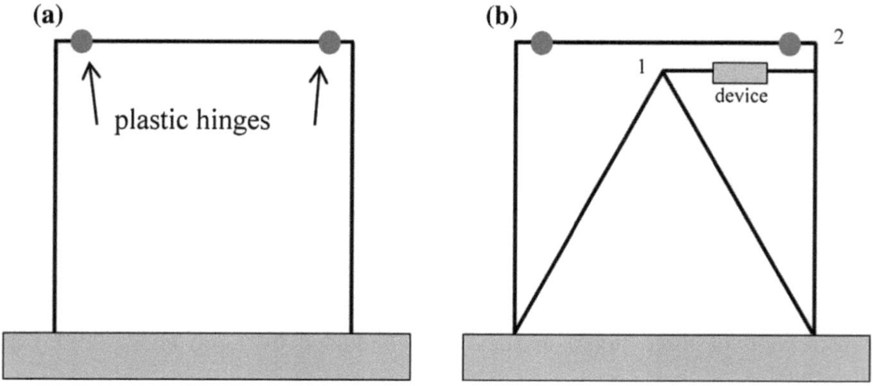

Fig. 1.13 Frame without and with passive energy dissipation devices (modified from Symans et al. 2008)

lateral strength of the frame is 0.2 times the weight of the frame, and the postyield stiffness is equal to 2.0 % of the initial stiffness. The vibration period of the structure, T_0, is 0.535 s and its inherent damping (in the absence of any passive energy dissipation device) is assumed to be 5 % of critical. The results from the non-linear response-history analyses of the bare frame, when it is subjected to the horizontal component of the earthquake ground motion reveals that plastic hinges form in the girder, the maximum drift is 1.03 % of the height of the structure and the corresponding displacement ductility demand is 3.08. At the end of the earthquake, the structure has a residual drift of 0.12 % of the story height. The damage in the frame can be quantified via a damage measure (DM) (Park 1984; Park et al. 1985, 1987; De Iuliis et al. 2010) such as that given by:

$$DM = \frac{\mu_{Demand}}{\mu_{Capacity}} + \rho \frac{E_{Demand}}{E_{Capacity}} \tag{1.12}$$

where

- μ_{Demand} and E_{Demand} are, respectively, maximum displacement ductility demand and cumulative hysteretic energy dissipation demand on the system or component;
- μ_{Capacity} and E_{Capacity} are, respectively, ductility capacity and hysteretic energy capacity for one full cycle of plastic deformation of the system or component;
- ρ is calibration factor.

The calibration factor (set equal to 0.15 for this example) is material dependent, and is selected to produce a damage measure value of 0.0 when the structure is undamaged, and 1.0 when the damage is severe (near or at incipient incipient collapse). Damage measure values in excess of 0.4 are generally considered unacceptable. For the bare frame shown in Fig. 1.13a, the value of DM is 0.955 and thus the bare frame is severely damaged. Note that a DM value of near 1.0 may be obtained by a single monotonic deformation demand that is equal to the deformation capacity, or as is most common, by undergoing numerous cycles of deformation demand that are significantly less than the deformation capacity.

The damage measure of Eq. (1.12) indicates that damage to the structure can be reduced by decreasing the ductility or hysteretic energy demand or by increasing the ductility or hysteretic energy capacity. Assuming that it is not economically feasible to increase the ductility or hysteretic energy capacity of the structure under consideration, the performance may only be improved by reducing the ductility or hysteretic energy dissipation demand. If a passive energy dissipation device in the form of a viscous fluid damper is used, the reduction in ductility demand is facilitated through displacement reductions that come with increased damping. When metallic yielding devices are utilized, the reduction in ductility demand is provided by reduced displacements that arise from increased stiffness of the system as well as hysteretic energy dissipation within the devices. In structures that employ passive energy dissipation devices, the hysteretic energy dissipation demand on critical components of the structure can be reduced by transferring the energy dissipation demand to the passive energy dissipation devices. For systems incorporating passive energy dissipation devices, it is useful to recast the viscous damping energy and hysteretic energy terms as follows:

$$E_D = E_{D,\,structure} + E_{D,\,devices} \tag{1.13}$$

$$E_H = E_{H,\,structure} + E_{H,\,devices} \tag{1.14}$$

the sum of the terms $E_{\text{D,devices}}$ and $E_{\text{H,devices}}$ gives the term E_p of the Eq. (1.6). In Eq. (1.13), the viscous damping energy is separated into damping that is inherent in the structure and added damping from passive energy dissipation devices. In Eq. (1.14), the first term represents the part of the hysteretic energy dissipated by the main structural and nonstructural elements, and the second part is that dissipated by the added passive energy dissipation devices.

To demonstrate the above principles, the structure of Fig. 1.13a is to be increased with a certain passive energy dissipation device, connected to the frame through a stiff chevron brace, as shown in Fig. 1.13b. One end of the device is attached to the top of the chevron brace and the other end is connected to the structure. The lateral stiffness of the brace, designed to remain elastic for all loadings, is approximately 4.0 times the initial lateral stiffness of the frame without the brace. The structure has two degrees of freedom (DOF); the lateral displacement of the top of the chevron brace and the lateral displacement of the top of the frame (numbered as one and two, respectively, on Fig. 1.13b). The device resists the relative motion (displacement and/or velocity) between these two points. Two different types of devices have been considered: a metallic yielding device and a viscous fluid device. The metallic device is referred to as a rate-independent device and the viscous device is classified as a rate-dependent device. The metallic device is rate independent since the resisting force in the device is a function only of the relative displacement across the device (i.e. the difference in displacements between DOF one and two). The viscous device is rate dependent since the resisting force in the device is dependent, in part or in full, on the relative velocity across the device (i.e. the difference in velocities between DOF one and two).

Seven different implementations of this device have been considered, where the yielding element has a strength of 0.167, 0.333, 0.500, 0.667, 0.833, 1.0, and 100 times that of the bare frame. The value of 100 represents a rigid connection between the chevron brace and the structure.

The elastic stiffness of the device increases with its strength since the device yield displacement has been assumed to be constant. The viscous damper has been a linear viscous fluid device that is implemented so that the total damping (inherent plus added) is 10, 15, 20, 25, 30, and 50 % of critical. In all of the analyses, the inherent damping of the structure is assumed to be 5 % of critical.

The performance of the structure with the added metallic yielding device is shown in Table 1.2. The elastic period of vibration of the structure, T, decreases with each increase in device strength, Y_A, and, correspondingly, elastic stiffness. Depending on the characteristics of the ground motion used for the analysis, this

Table 1.2 Effect of added metallic yielding device on structure performance (modified from Symans et al. 2008)

Y_A/Y_O	T_A/T_O	Δ_{max}	ΔR_{max}	V_B/W_O	$\mu_{Demand}/\mu_{Capacity}$	$E_{Demand}/E_{Capacity}$	DM
0	1.000	0.01027	0.00117	0.223	0.513	0.736	0.955
0.167	0.869	0.01033	0.00097	0.261	0.517	0.520	0.829
0.333	0.796	0.00867	0.00182	0.296	0.433	0.327	0.629
0.500	0.751	0.00747	0.00141	0.319	0.373	0.213	0.501
0.667	0.720	0.00645	0.00253	0.349	0.323	0.143	0.409
0.833	0.695	0.00707	0.00269	0.384	0.353	0.099	0.413
1.000	0.679	0.00707	0.00189	0.424	0.353	0.076	0.399
100	0.523	0.00364	0.00013	0.685	0.182	0.017	0.192

decrease in period may be responsible for increasing base shear in the structure. Although the drift ratio, Δ_{max}, decreases significantly with increased device capacity, the residual deformation of the structure, $\Delta\ R_{max}$, is increased in most cases due to the residual plastic deformation in the metallic yielding device.

The residual deformation is not necessarily a concern since the devices can be replaced after an earthquake. The base shear demand, V_B, increases significantly with increased device capacity, and is nearly doubled when the device strength is equal to the original strength of the structure. The increased base shear would need to be accommodated in the design of the structure and its foundation. The ductility and energy dissipation demands and, correspondingly, the damage measure, decrease significantly with each increase in device capacity.

When the device strength is equal to the original strength of the structure, DM is reduced to 0.399, which is at the upper limit of acceptability. Although DM is reduced further to 0.192 when the brace is rigidly connected to the structure (i.e. without a device), the base shear is increased by a factor of more than 3.0, which may not be acceptable.

The performance of the structure with the added linear viscous fluid damping device is shown in Table 1.3.

The elastic period of vibration of the structure, *T*, does not change with added viscous damping. This is because viscous damping devices have zero or negligible stiffness under low-frequency response. The drift ratio decreases by about 50 %

Table 1.3 Effect of added viscous fluid damping device on structure performance (modified from Symans et al. 2008)

ξ	T/T_O	Δ_{max}	ΔR_{max}	V_B/W_O	$\mu_{Demand}/\mu_{Capacity}$	$E_{Demand}/E_{Capacity}$	DM
0.05	1.00	0.01027	0.00117	0.223	0.513	0.736	0.955
0.10	1.00	0.00940	0.00175	0.264	0.470	0.494	0.767
0.15	1.00	0.00847	0.00187	0.293	0.423	0.350	0.633
0.20	1.00	0.00767	0.00177	0.312	0.383	0.250	0.34
0.25	1.00	0.00700	0.00052	0.324	0.350	0.185	0.461
0.30	1.00	0.00635	0.00001	0.333	0.317	0.139	0.401
0.50	1.00	0.00517	0.00118	0.351	0.259	0.049	0.288

when the total viscous damping ratio is increased from the inherent level of 5 % to a total of 50 % (i.e. 45 % added damping). The residual deformation of the structure is affected by the implementation of the device, but there is no distinctive trend. The base shear demand, V_B, increases significantly with increased damping. This is due to the fact that the structure is behaving inelastically and the damping force increases linearly with both the damping coefficient and the velocity and, for this particular example structure, the damping coefficient increases faster than the velocity decreases. Sadek et al. (2000) showed that increases in base shear can also occur for elastic structures, particularly for structures with long natural periods.

Finally, it is noted that the damage measure, DM, has decreased from 0.955 for the structure without the device, to 0.4 for the structure with a total damping ratio of 30 %. Although the DM is decreased even further for the system with 50 % damping, it may be impractical to achieve that much added damping at a reasonable cost. It is also important to note that, even with a total damping ratio of 50 %, the main structural system still yields. Experience has shown that, for strong earthquakes, it is virtually impossible to add enough damping to completely avoid yielding (and hence, damage) in the structural framing system (Uriz and Whittaker 2001; Oesterle 2003).

In summary, both the metallic yielding and viscous fluid damping devices were highly effective in reducing damage in the structure. However, with particular reference to the metallic yielding devices, this comes at the expense of increased base shear and therefore foundation costs.

References

Applied Technology Council (ATC): Review of Seismic Research Results on Existing Buildings. ATC-37, Redwood City, California (1994)

Constantinou, M.C., Symans, M.D.: Seismic response of structures with supplemental damping. Struct. Des. Tall Build. **22**, 77–92 (1993)

Constantinou, M.C., Soong, T.T., Dargush, G.F.: Passive Energy Dissipation Systems for Structural Design and Retrofit. Monograph of the National Center for Earthquake Engineering Research, Buffalo (1996)

De Iuliis, M., Castaldo, P.: An energy-based approach to the seismic control of one-way asymmetrical structural systems using semi-active devices. Ingegneria Sismica, Patron Editore, Anno **XXIX**(4), 31–42 (2012)

De Iuliis, M., Castaldo, P., Palazzo, B.: Analisi della domanda sismica inelastica del terremoto de L'Aquila su sistemi dimensionati secondo le NTC2008. Ingegneria Sismica, Patron Editore **3**, 52–65 (2010)

Fujino, Y., Soong, T.T., Spencer, B.F. Jr.: Structural control: basic concepts and applications. In: Proceedings of ASCE Structures Congress XIV, pp. 1277–1287. Chicago, Illinois (1996)

Kelly, J.M.: Earthquake-Resistant Design with Rubber, 2nd ed. Berlin and New York: Springer-Verlag (1997)

Lobo, R.F., Bracci, J.M., Shen, K.L., Reinhorn, A.M., Soong, T.T.: Inelastic response of R/C structures with viscoelastic braces. Earthq. spectr. **9**(3), 419–446 (1993)

Oesterle, M.: Use of incremental dynamic analysis to assess the performance of steel moment-resisting frames with fluid viscous dampers. MS Thesis, Virginia Tech, Blacksburg (2003)

Park, Y.J.: Seismic damage analysis and damage-limiting design of R/C structures. Ph.D Thesis, Department of Civil Engineering, University of Illinois, Urbana, (1984)

Park, Y.J., Ang, A.H.S., Wen, Y.K.: Seismic damage analysis of reinforced concrete buildings. J. Struct. Eng. **111**(4), 740–757 (1985)

Park, Y.J., Ang, A.H.S., Wen, Y.K.: Damage-limiting aseismic design of buildings. Earthq. Spectr. **3**(1), 1–26 (1987)

Sadek, F., Mohraz, B., Riley, M.A.: Linear procedures for structures with velocity-dependent dampers. J. Struct. Eng. **126**(8), 887–895 (2000)

Soong, T.T., Constantinou, M.C.: Passive and Active Structural Vibration Control in Civil Engineering. Springer, Wien (1994)

Soong, T.T., Dargush, G.F.: Passive Energy Dissipation Systems in Structural Engineering. Wiley and Sons Ltd, Chichester (1997)

Soong, T.T., Dargush, G.F.: Passive energy dissipation and active control, Structural Engineering Handbook. CRC Press LLC, Boca Raton (1999)

Soong, T.T., Spencer, B.F. Jr.: Supplemental energy dissipation: state of the art and state of the practice. Eng. Struct. **24,** 243–259 (2002)

Symans, M.D., Constantinou, M.C.: Seismic testing of a building structure with a semi-active fluid damper control system. Earthq. Eng. Struct. Dynam. **26**(7), 759–777 (1997)

Symans, M.D., Constantinou, M.C.: Semi-active control systems for seismic protection of structures: a state of the art review. Eng. Struct. **21**, 469–487 (1999)

Symans, M.D., Charney, F.A., Whittaker, A.S., Constantinou, M.C., Kicher, C.A., Johnson, M.W., McNamara, R.J.: Energy dissipation system for seismic applications: current practise and recent developments. J. Struct. Eng. **134**(1), 3–21 (2008)

Uriz, P., Whittaker, A.S.: Retrofit of prenorthridge steel moment-resisting frames using fluid viscous dampers. Struct. Des. Tall Build. **19**(5), 371–390 (2001)

Whittaker, A.S., Aiken, I.D., Bergman, D., Clark, P.W., Cohen, J., Kelly, J.M., Scholl, R.E.: Code requirements for design and implementation of passive energy dissipation systems. In: Proceedings of ATC-17–1 Seminar on Seismic Isolation, Passive Energy Dissipation and Active Control, vol. 2, pp. 497–508. ATC, Redwood City, California (1993)

Chapter 2
Passive Energy Dissipation Devices

Abstract In this chapter, the supplemental passive energy dissipation devices, metallic or hysteretic dampers, frictional, viscoelastic and viscous (linear and non-linear) dampers, developed and studied over the years, are briefly described. For each type of device, the physical, mechanical and technological aspects are analysed by describing the construction, hysteretic behavior, physical models, advantages, and disadvantages. Then, the more appropriate mathematical laws to model their dynamic behavior, with particular reference to the viscous and viscoelastic, are described. Finally, a comparison between all the different types of device is reviewed and the main recommendations, reported in the international codes with specific reference to the viscous and viscoelastic, are explained.

2.1 Dissipation Devices

Several passive energy dissipation devices are available and have been implemented worldwide for the seismic protection of structures.

The passive energy dissipation devices can be classified into three categories:

(1) rate-dependent devices;
(2) rate-independent devices;
(3) others.

The rate-dependent devices are the dampers whose mechanical response depends on the relative velocity between the ends of the device. The behavior of these dampers is commonly described using different models of linear viscoelasticity. Examples of these include viscous fluid or viscoelastic dampers and viscoelastic solid dampers.

The viscoelastic fluid dampers, in general, have a dynamic behavior characterized by low stiffness values over a range of frequencies which often includes the fundamental natural frequency of buildings or bridges. Therefore, these devices

P. Castaldo, *Integrated Seismic Design of Structure and Control Systems*,
Springer Tracts in Mechanical Engineering, DOI: 10.1007/978-3-319-02615-2_2,

generally have a minimal influence on the fundamental natural frequency and, therefore, are often regarded simply as viscous fluid devices.

The viscoelastic solid dampers have a stiffness that influences the natural frequencies of vibration of the structure.

The rate-independent systems are the dampers whose mechanical response depends on the displacement between the ends of the device and the sign of the velocity (i.e., the direction of motion). The behavior of these devices is commonly described by using various non-linear hysteretic models. Typical examples include metallic and friction devices. The metallic devices have a good hysteretic behavior associated with the yield point of mild steel, while the friction ones have a fundamentally bilinear hysteretic behavior with a very high initial stiffness.

In the following, the hysteretic (metallic), friction, viscous fluid and viscoelastic devices will be discussed.

2.2 Metallic Dampers

One of the most effective mechanisms available for the dissipation of energy, input to a structure during an earthquake, is through the inelastic deformation of metallic substances (Soong and Dargush 1997).

The idea of using separate metallic hysteretic dampers within a structure to absorb a large portion of the seismic energy began with the conceptual and experimental work by Kelly et al. (1972) and Skinner et al. (1975). Several of the devices considered by these researchers included torsional beam, flexural beam, and U-strip dampers as shown schematically in Fig. 2.1.

Another type of device is the buckling restrained brace (BRB), developed by Wada in 1980 (Wada et al. 1998), shown in Fig. 2.2.

A BRB damper consists of a steel brace (usually having a low-yield strength) with a cruciform cross section that is surrounded by a stiff steel tube (Symans et al. 2008). The region between the tube and brace is filled with a concrete-like material

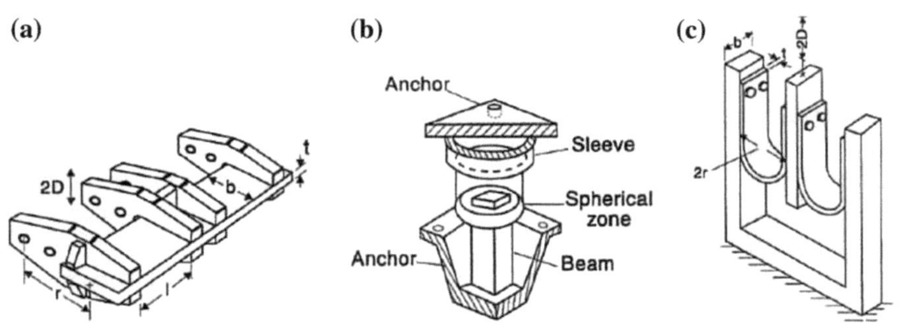

Fig. 2.1 Metallic damper geometries **a** torsional beam; **b** flexural beam; **c** U-strip (modified from Soong and Dargush 1997)

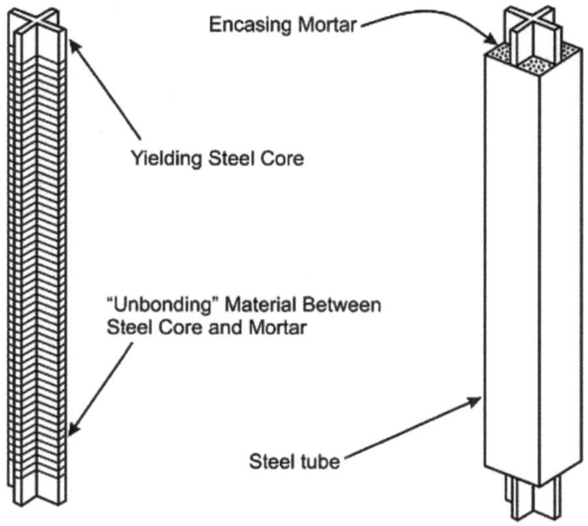

Fig. 2.2 Buckling restrained brace

and a special coating is applied to the brace in order to prevent it from bonding to the concrete. Thus, the brace can slide with respect to the concrete-filled tube. The confinement provided by the concrete-filled tube allows the brace to be subjected to compressive loads without buckling (i.e., the damper can yield in tension or compression with the tensile and compressive loads being carried entirely by the steel brace). Under compressive loads, the damper behavior is essentially identical to its behavior in tension. Since buckling is prevented, significant energy dissipation can occur over a cycle of motion. In many cases, BRB dampers are installed within a chevron bracing arrangement. In this case, under lateral load, one damper is in compression and the other is in tension, and hence zero vertical load is applied at the intersection point between the dampers and the beam above. Thus, the dampers may be regarded as superior to a conventional chevron bracing arrangement where the compression member is expected to buckle elastically, leaving a potentially large unbalanced vertical force component in the tension member that is, in turn, applied to the beam above. During the initial elastic response of the BRB damper, the device provides stiffness only. As the BRB damper yields, the stiffness reduces and energy dissipation occurs due to the inelastic hysteretic response. A mathematical model that describes the hysteretic behavior of metals is the Bouc–Wen model (Wen 1976), which is described by Black et al. (2004). The model is defined by:

$$P(t) = \beta K u(t) + (1 - \beta) K u_y Z(t) \tag{2.1}$$

where β is the ratio of post- to preyielding stiffness, K is the preyielding stiffness, u_y is the yield displacement, and Z(t) is the evolutionary variable that is defined by:

$$u_y \dot{Z}(t) + \gamma |\dot{u}(t)||Z(t)||Z(t)|^{\delta-1} + \eta \dot{u}(t)|Z(t)|^{\delta} - \dot{u}(t) = 0 \tag{2.2}$$

where γ, δ and η are dimensionless parameters that define the shape of the hysteresis loop. For example, for large values of δ, the transition from elastic to inelastic behavior is sharp and the hysteresis loop is associated with a bilinear model.

In order to effectively include these devices in the design of an actual structure, their expected non-linear force–displacement behavior under arbitrary cyclic loads must be characterized. Özdemir (1976) was the first to consider this modeling problem. He used analogies with existing elastoplastic and viscoplastic constitutive theories to develop appropriate forms for the force–displacement relationships.

Over the ensuing years, considerable progress has been made in the development of metallic dampers. For example, many new designs have been proposed, including the X-shaped and triangular plate dampers, known as added damping and stiffness (ADAS), reported in Fig. 2.3.

These ADAS dampers (Whittaker et al. 1989, Whittaker et al. 1991; Xia and Hanson 1992; Fierro and Perry 1993; Aiken et al. 1993) consist of a series of steel plates wherein the bottom of the plates are attached to the top of a chevron bracing arrangement and the top of the plates are attached to the floor level above the bracing. As the floor level above deforms laterally with respect to the chevron bracing, the steel plates are subjected to a shear force. The shear forces induce bending moments over the height of the plates, with bending occurring at the weak axis of the plate cross section. The geometrical configuration of the plates is such that the bending moments produce a uniform flexural stress distribution over the height of the plates. Thus, inelastic action occurs uniformly over the full height of the plates. For example, in the case where the plates are fixed-pinned, the geometry is triangular, in the case where the plates are fixed–fixed, the geometry is an hourglass shape, as shown in Fig. 2.3. To ensure that the relative deformation of the ADAS device is approximately equal to that of the story in which it is installed, the chevron bracing must be very stiff.

Alternative materials, such as lead and shape-memory alloys, have been evaluated. Numerous experimental investigations have been conducted to determine performance characteristics of the individual devices and laboratory test structures. As a result of this ongoing research program, several commercial products have been developed and implemented in both new and retrofit construction projects (Soong and Dargush 1997). In particular, a number of existing structures in New Zealand, Mexico, Japan, Italy, and the United States now include metallic dampers as a means for obtaining improved seismic resistance.

2.2.1 Costitutive Models

To design hysteretic devices and implement global analyses of structures equipped with these dampers, the first step is to define the constitutive model of the metal used (Soong and Dargush 1997).

Consider first the behavior of a cylindrical metal rod, with initial length L_0 and cross-sectional area A_0, subjected to uniaxial tension as shown in Fig. 2.4.

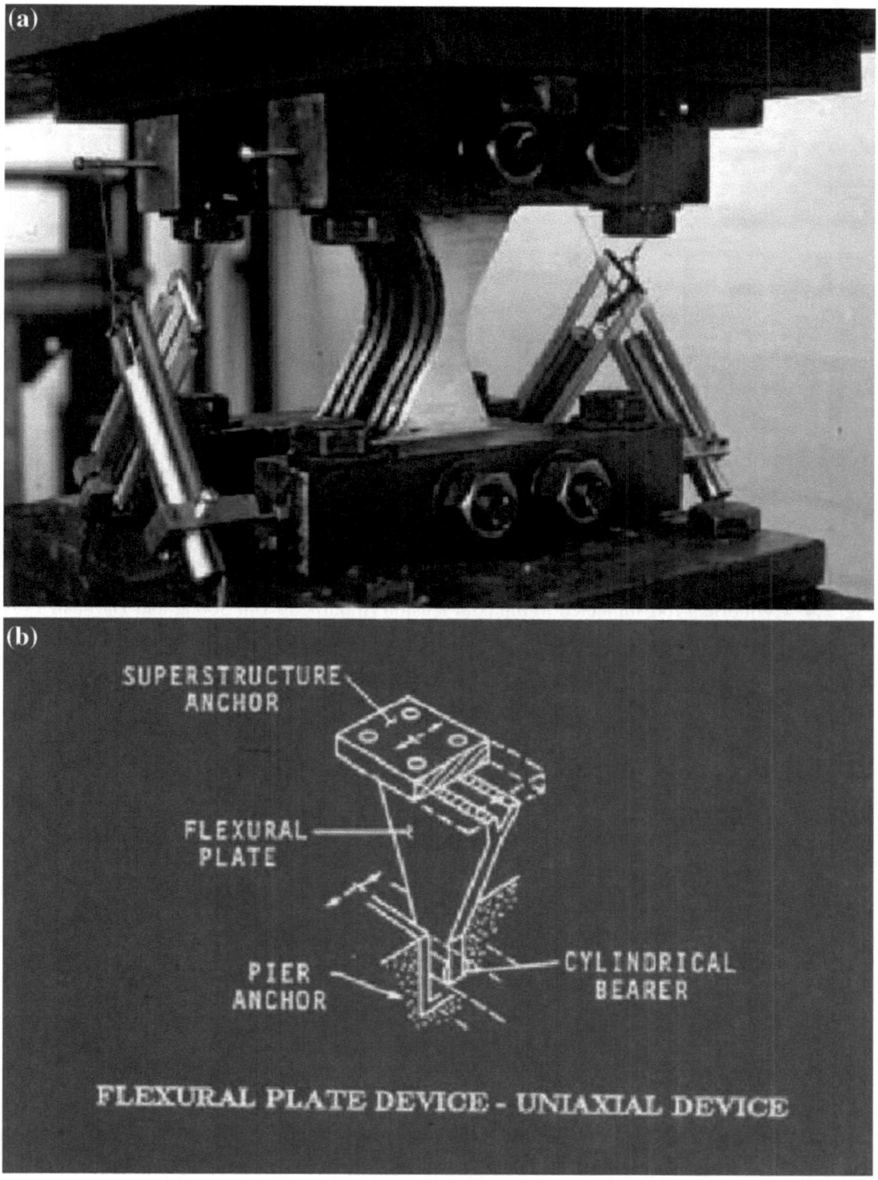

Fig. 2.3 Metallic dampers

Assuming that the load P is incremented slowly to insure the validity of a quasistatic approximation, typical stress–strain curves are displayed in Fig. 2.5.

In these diagrams, the abscissa represents the conventional strain ε, while the nominal stress σ_n is plotted on the ordinate-axis. The following relationships apply:

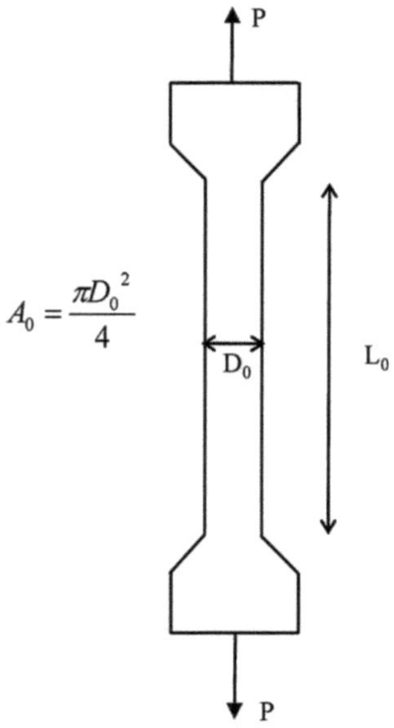

Fig. 2.4 Cylindrical rod subjected to uniaxial tension (modified from Soong and Dargush 1997)

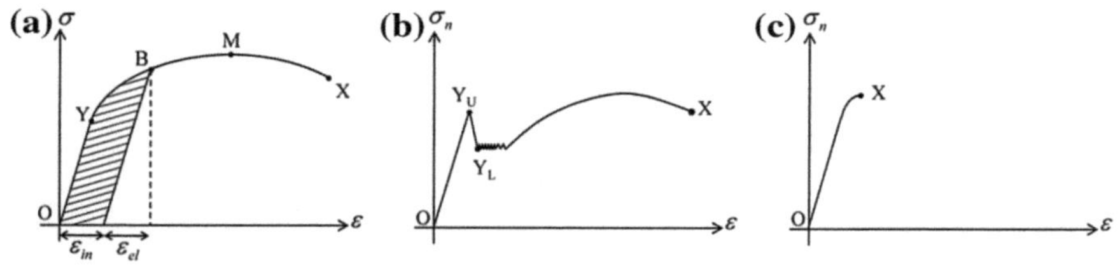

Fig. 2.5 Stress–strain curves

$$\varepsilon = \frac{L - L_0}{L_0} \tag{2.3}$$

$$\sigma_n = \frac{P}{A_0} \tag{2.4}$$

The curve in Fig. 2.5a is characteristic of most metals. At loads corresponding to nominal stress less than the yield stress σ_y, the response is fully elastic with σ_n proportional to ε. In this range, the initial state O is fully recoverable with removal of the applied load, and there is no energy dissipation. On the other hand, when the nominal stress exceeds the yield stress (i.e., beyond point Y on the curve), irreversible plastic deformation occurs dissipating energy. Considering the state labeled B, it is useful to partition the total strain at B into elastic (ε_{el}) and inelastic (ε_{in}) contributions, as indicated in Fig. 2.5a. Thus:

$$\varepsilon = \varepsilon_{el} + \varepsilon_{in} \tag{2.5}$$

in which:

$$\varepsilon_{el} = \frac{\sigma}{E} \tag{2.6}$$

with E representing the elastic modulus. The energy, or more precisely the energy density, is measured by the area under the stress–strain curve from O to B. Part of that energy is recoverable. However, the remainder, associated with the inelastic strain ε_{in} and identified by the shaded area, is dissipative. A significant portion of that dissipative energy is converted into heat. In this strain hardening regime, successive increments of stress produce correspondingly greater increments of strain, until the point M is reached associated with the maximum load that the tensile specimen can withstand. Beyond M, the specimen becomes unstable. Localization phenomena start with the appearance of a necking region in which three-dimensional states of stress are present. Ultimately, failure occurs at point X.

As noted above, this description applies to most metals. However, annealed mild steel and some other alloys behave as illustrated in Fig. 2.5b. In this case, the response is similar to that previously discussed, except in the region just beyond first yield at Y_U. In these materials, there is an abrupt drop in stress from point Y_U to Y_L, corresponding to the upper and lower yield stress, respectively. This phenomenon and the ensuing stress–strain plateau is caused by the formation and propagation of Lüders bands (also called *Hartmann lines*). It should be noted, however, that strain hardened mild steel does not exhibit this behavior, but instead follows that depicted in Fig. 2.5a (Soong and Dargush 1997).

The stress–strain curve, shown in Fig. 2.5c, is typical of a brittle metal, such as cast iron. Obviously, this type of material is not a good candidate for use in metallic dampers, since very little energy dissipation occurs before fracture.

The definitions used in Eqs. (2.3) and (2.4) are appropriate when the length and area of the specimen do not change significantly from the initial values. However, for higher strain levels, the natural strain e and true stress σ are more appropriate measures, where:

$$e = \ln\left(\frac{L}{L_0}\right) \tag{2.7}$$

$$\sigma = \frac{P}{A} \tag{2.8}$$

Replotting Fig. 2.5a in these coordinates leads to the curve displayed in Fig. 2.6.

Over the years, several mathematical models have been introduced to idealize the stress–strain curves, including the elastic-perfectly plastic model, the elastic-linear strain hardening model, and the Ramberg–Osgood model shown in Fig. 2.7 (Soong and Dargush 1997). The last of these models can be written as:

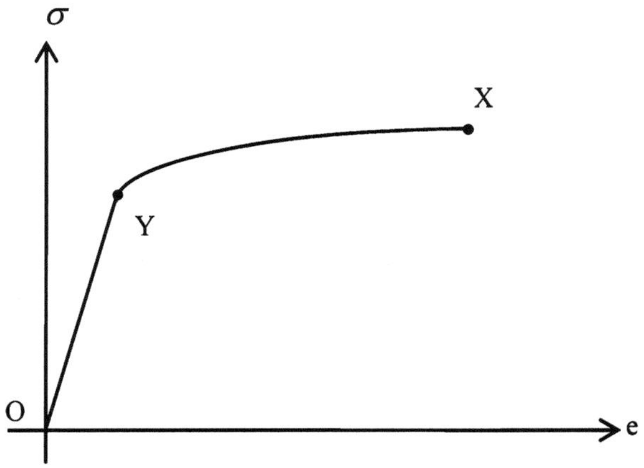

Fig. 2.6 Nominal stress–conventional strain diagrams

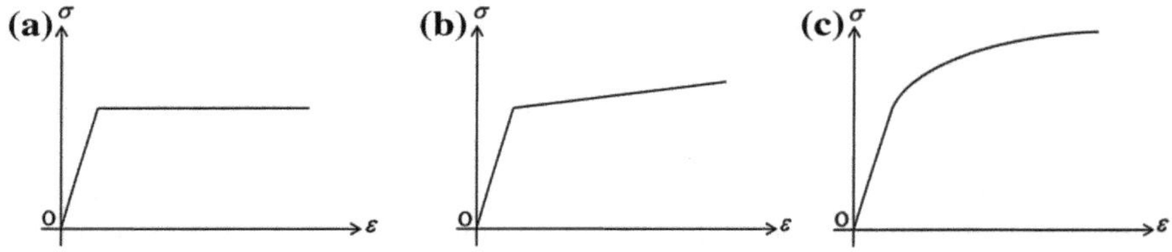

Fig. 2.7 Stress–strain mathematical models: **a** elastic-perfectly plastic; **b** elastic-linear strain hardening; **c** Ramberg–Osgood

$$\varepsilon = \frac{\sigma}{E} + k\left(\frac{\sigma}{E}\right)^n \tag{2.9}$$

with material constants E, k, and n. Thus, the Ramberg–Osgood model essentially establishes a power law relationship between stress and inelastic strain, a condition which is approximately satisfied during monotonic loading experiments with different metals.

However, Eq. (2.9) is not adequate for describing the response to arbitrary cyclic loading in which the stress state depends on not only the current strain but rather the entire prior history. Consequently, in order to develop models for metallic dampers, the discussion must be extended to behavior under load reversals that involve excursions into the inelastic range. For these cases, the response is path dependent. Consider the cylindrical specimen first loaded in tension past yield at Y to a point B in the strain hardening range, and then gradually unloaded as indicated in Fig. 2.8 (Soong and Dargush 1997).

The unloading branch of the curve is parallel to the initial loading curve, indicating a purely elastic response. However, if unloading continues sufficiently into the compression range, yielding will again take place at the point labeled Y′.

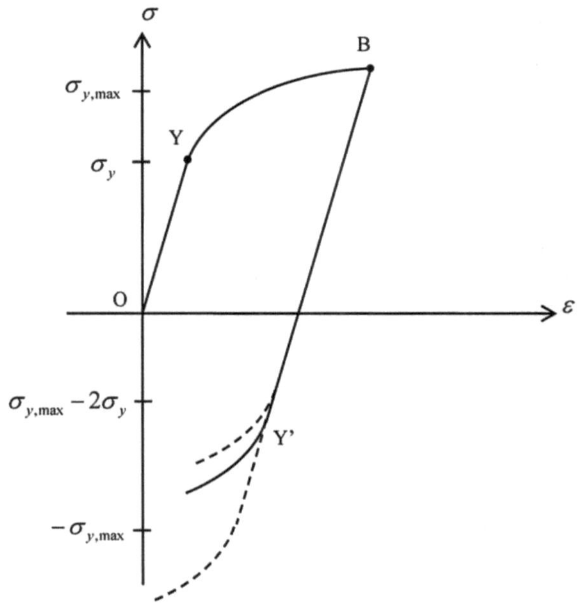

Fig. 2.8 Cyclic stress–strain response (modified from Soong and Dargush 1997)

The stress associated with Y′, i.e. σ_y, depends on the prior amount of strain hardening. This is known as the Bauschinger effect.

In all the above-mentioned formulations of the constitutive models, time dependency has been ignored. Plastic flow is assumed to occur instantaneously compared to the time variation of the applied load. This is reasonable for steel at approximately room temperature deforming under moderate strain rates. It is not appropriate for lead under similar conditions, nor for steel at high temperature or under very high strain rates. In the latter cases, creep and relaxation phenomena must be considered. Creep is characterized by increasing strain with time for a constant stress, while relaxation signifies a continual reduction in stress with time for a material under constant strain. A volume edited by Miller (1987) provides an excellent introduction to the unified creep-plasticity state-variable models. However, Ozdemir (1976) developed a particular formulation for the examination of metallic dampers.

The theories of plasticity and viscoplasticity are able to characterize the behavior of metals in the inelastic range during cyclic loading. However, these models do not typically include the notion of failure. However, it is known that metals subjected to cycling loads will often fail due to fatigue. The phenomenon of low-cycle fatigue of metallic dampers is of particular interest, with it resulting in a limited number of excursions (e.g. <1,000) well into the inelastic range. The actual mechanism involves the growth and interconnection of microcracks that eventually lead to failure at the macroscopic level. Since this process is very difficult to trace, a more phenomenological approach is generally adopted based upon the concept of material damage. The traditional approach for low-cycle fatigue analysis is based on the results of uni-axial tests at various levels of cyclic deformation (Hertzberg 1983) and refers to the rule of Palmgren-Minor cumulative damage.

2.2.2 Damper Modeling

Having discussed the mechanical behavior of metals at the constitutive level, in this section the emphasis now shifts to the characterization of the overall metallic damper response. In particular, the construction of the force–displacement model is described from an appropriate constitutive relationship for the metal by applying the principles of mechanics.

Consider the triangular plate metallic damper (TADAS), which consists of N identical triangular structural steel plates positioned in parallel and is typically installed within a frame bay between a chevron brace and the overlying beam. The base of each triangular plate is welded into a rigid base plate to approximate a fixed end condition, while a slotted pin connection is employed at the apex to ensure relatively free movement in the vertical direction. As a result of this configuration, the damper primarily resists horizontal forces P, associated with an interstory drift D, via uniform flexural deformation of the individual plates. Thus, it is appropriate to examine a single cantilevered plate of thickness h, length L, and base width w_0, subjected to a load P/N applied at its free end as detailed in Fig. 2.9.

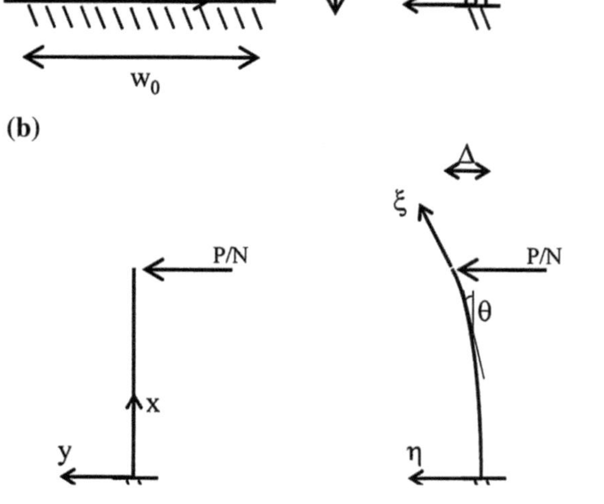

Fig. 2.9 Mathematical model for triangular plate device: **a** geometric definition; **b** beam idealization (modified from Soong and Dargush 1997)

The force–displacement relationship for the damper can be readily established for an infinitesimal elastic response. In this case, the classical Euler–Bernoulli beam theory is valid. A quasistatic formulation is adopted by ignoring the inertia of the plate. Then, at any cross-section, the moment equilibrium equation can be written:

$$\frac{P}{N}(L-x) = \frac{w_0(L-x)}{L}\int_{-\frac{h}{2}}^{+\frac{h}{2}} \sigma y dy \tag{2.10}$$

It is evident that the stress is independent of the position along the beam axis. After applying the unidirectional elastic constitutive relationship:

$$\sigma = E\varepsilon \tag{2.11}$$

and the kinematic condition:

$$\varepsilon = ky \tag{2.12}$$

the curvature k is constant along the entire length of the plate, with:

$$k = \frac{2\Delta}{L^2} \tag{2.13}$$

After some arithmetic operations and substitutions, the following force–displacement model can be obtained:

$$P = \left(\frac{NEw_0h^3}{6L^3}\right)\Delta \tag{2.14}$$

The above model pertains only to the elastic response of the damper. Very little can be inferred concerning the energy dissipation characteristics of the device. This type of information, which is vital for proper aseismic design, requires examination of the inelastic response. As a first approximation, the Eqs. (2.12) and (2.13) can be retained, while replacing Eq. (2.11) by a rate-independent inelastic constitutive model. Under certain conditions (e.g. major earthquakes), the interstory drift D may become comparable in magnitude to the damper plate length L. In these situations, the effects of finite deformation on damper response cannot be ignored. Thus, it is necessary to write the moment equilibrium equation in the deformed configuration. With finite deformation, the curvature k is no longer constant along the length of the plate. Consequently, at each instant of time, a nonlinear boundary value problem (BVP) must be solved to determine the unknown curvature function.

2.3 Friction Dampers

Friction dampers utilize the mechanism of solid fiction that develops between two solid bodies sliding relative to one another to provide the desired energy dissipation (Soong and Dargush 1997). Processes of this type are prevalent in nature and have also been employed in many engineering systems. For example, solid friction plays an important role in the overall control of tectonic movement and earthquake generation. On a much smaller scale, friction is also used in automotive brakes as a means to dissipate the kinetic energy of motion. Based primarily upon an analogy to the automotive brake, Pall et al. (1980) began the development of passive frictional dampers to improve the seismic response of structures. The objective is to slow down the motion of buildings "by braking rather than breaking" (Pall and Marsh 1982).

The friction devices exhibit a hysteretic behavior similar to that shown by the metallic devices. They are based on the resistance developed between two interfaces in motion in order to dissipate an amount of input energy in the form of heat. During the seismic excitations, the device flows under a predetermined load, providing the desired energy dissipation. The friction device, which are not susceptible to thermal effects, have a reliable performance as well as a stable hysteretic behavior.

Over recent years considerable progress has been made in the development of friction devices. Some typologies are shown in Fig. 2.10.

The Limited Slip Bolted (LSB) joint proposed by Pall et al. (1980) is shown in Fig. 2.10a. It is intended for seismic control of large panel structures. The LSB design incorporated brake lining pads between steel plates in order to provide a consistent force–displacement response. Figure 2.10b displays an alternative design proposed by Pall and Marsh (1982) for application in conjunction with cross-bracing in framed structures. Once again, brake lining pads are utilized for the sliding surfaces. Modern versions of these devices have already been implemented in a number of structures in Canada. Two more recent uniaxial friction devices are shown in Fig. 2.10c and d. The first of these is a Sumitomo friction damper that has been applied in Japan (Aiken and Kelly 1990). The copper alloy friction pads slide along the inner surface of the cylindrical steel casing. The required normal force is provided through the action of the spring against the inner and outer wedges. The device can be installed in-line with diagonal bracing or in parallel to the beams of the connection between the floor and the V brace configuration. In general the displacements are reduced compared to the original configuration of the frame, the shear forces at the base and the accelerations are modified very little, and in some cases increased due to the increased stiffness of the new configuration.

Figure 2.10d presents the somewhat more sophisticated Energy Dissipating Restraint (EDR) described in Nims et al. (1993). In this design, dissipation occurs on the interface between the bronze friction wedges and steel cylinder wall. The

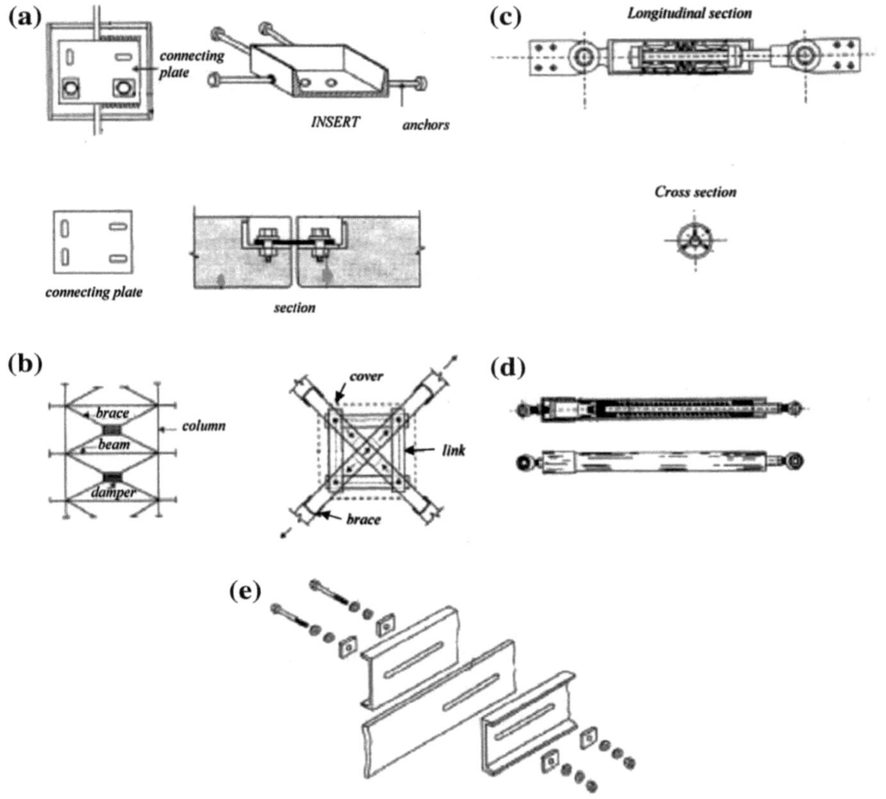

Fig. 2.10 Friction dampers: **a** limited slip bolt joint; **b** X-braced friction damper; **c** Sumitomo friction damper; **d** energy dissipating restraint; **e** slotted bolted connection (modified from Soong and Dargush 1997)

combination of wedges, stops, and internal spring produces a frictional force proportional to the relative displacement between the device ends.

Meanwhile, Fig. 2.10e shows a typical Slotted Bolted Connection (SBC), which provides a satisfactory performance as an energy dissipation element and even under repeated cycles of displacement does not present strenght loss, problems of stability and reduction of the dissipation capacity (Butterworth and Clifton 2000). Important factors that affect the performance of the SBC include:

- maintaining the contact pressure between the sliding surfaces;
- maintaining a more or less constant coefficient friction between the surfaces;
- avoiding brittle fracture of each component of the joint when it reaches the limit of the range of sliding;
- simple and inexpensive construction and maintenance.

Several versions have been discussed in literature. FitzGerald et al. (1989) employed all structural steel components, while Grigorian et al. (1993) advocated the inclusion of brass insert plates. In both cases, Belleville washers have been used to maintain initial bolt tensions.

The concentric bracing are among the most efficient systems to absorb the lateral actions, combining strength, stiffness, low weight and simplicity of realization. Unfortunately, the use of diagonal in tension leads to poor performance with potential soft-story collapses due to permanent strain. A better alternative consists of the use of sliding joints to protect the diagonals from damage. The SBCs (SBJs) are commonly used in a braced frame positioning them at one end of each diagonal and designed so as to slide before the yield strength or the instability of the diagonal. They were also used in a single brace but the need of capacity in compression of the SBC leads to major sections. A K configuration still requires braces able to withstand compression, but they will be shorter and lighter than the ones of a standard bracing as shown in Fig. 2.11.

2.3.1 Damper Behavior and Modeling

The scientific study about dry friction (contacting surfaces must remain dry during the process to maximize dissipation, there is no need for lubricants) has a long history that dates backs to the illustrious work of Leonardo da Vinci, Amontons, and Coulomb. The basic theory is founded upon the following hypotheses, which were initially inferred from physical experiments involving planar sliding of rectilinear blocks:

1. The total frictional force that can be developed is independent of the apparent surface area of contact.
2. The total frictional force that can be developed is proportional to the total normal force acting across the interface.

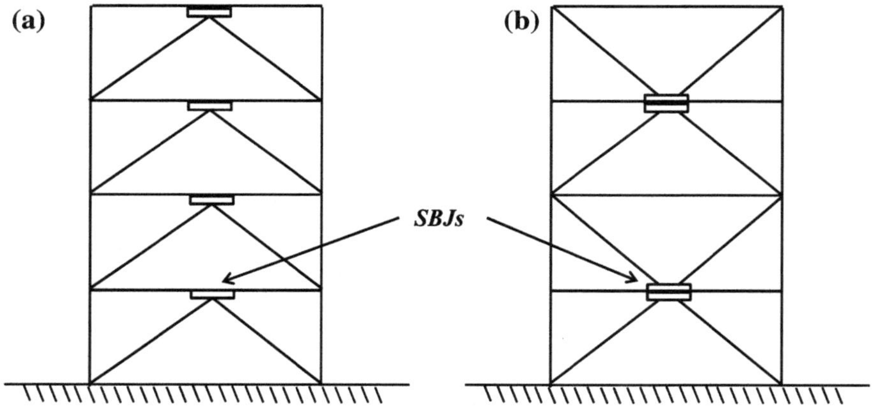

Fig. 2.11 K and X configurations: **a** K-braced; **b** X-braced

3. For the case of sliding with low relative velocities, the total frictional force is independent of that velocity.

As a result of these assumptions, at the instant of impending slippage or during sliding itself, it is possible to write:

$$F = \mu N \tag{2.15}$$

where F and N represent the frictional and normal forces, respectively, and μ is the coefficient of friction. Since it is frequently observed that the coefficient of friction is somewhat higher when slippage is imminent than it is during sliding, separate static (μ_s) and kinetic (μ_d) coefficients are often introduced. In either case, the frictional force F acts tangentially within the interfacial plane in the direction opposing the motion or impending motion.

In order to extend the theory to more general conditions, involving non-uniform distributions or non-planar surfaces, these basic assumptions are often abstracted to the infinitesimal limit. Thus, total forces are replaced by surface tractions, and the generalization of Eq. (2.15) becomes:

$$\tau_t = \mu \tau_n \tag{2.16}$$

in terms of the tangential and normal tractions. In practice, the Coulomb theory is only approximately true.

Furthermore, although the coefficient of friction μ is often assumed to be a constant for a given pair of contacting materials, this is not always the case. For example, the value of μ at any instant depends not only upon the selection of sliding materials, but also on the present condition of the sliding interface. Since surfaces are often the site of numerous ongoing physical and chemical processes, the friction coefficient associated with an interface may actually vary considerably over time.

It should consequently be clear that the development of force–displacement models for friction dampers must depend on the results of physical testing (Soong and Dargush 1997) in order to determinate both the most appropriate mathematical model and value of the friction coefficient.

Figure 2.12 details the hysteretic behavior of a simple brake lining frictional system under a constant amplitude displacement-controlled cyclic loading.

Based upon the behavior obtained by Pall et al. (1980), characterization of their simple brake lining frictional system in terms of an elastic-perfectly plastic model is quite appropriate. The model can be employed to simulate the behavior of their Limited Slip Bolted (LSB) joint is shown in Fig. 2.13a and b.

Notice that the model sketched in the diagrams includes a stiff bearing stage for displacements beyond the slip length. In order to quantify the response, a mathematical model must also be provided. A suitable hysteretic model can be defined, where the parameters P_s and Δ_b represent the slip load and total displacement at first contact with the bearing surface, respectively. Additionally, K_0 is the elastic

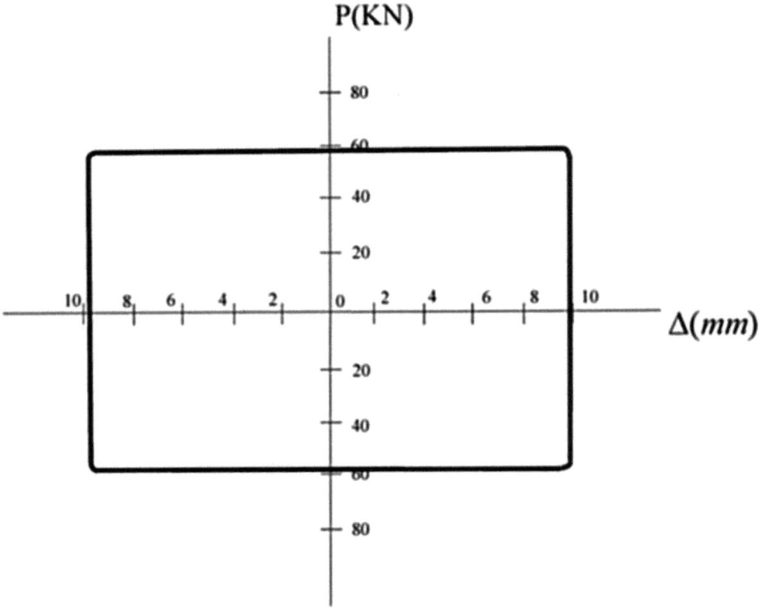

Fig. 2.12 Hysteresis loops of limited slip bolted joints (modified from Soong and Dargush 1997)

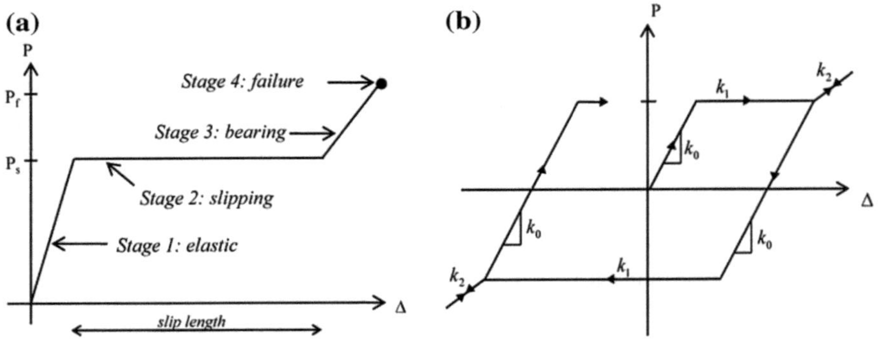

Fig. 2.13 Macroscopic model for limited slip bolted joints: **a** Load-deformation; **b** Hysteretic behavior

stiffness, while the stiffness in bearing is K_2. The model, which assumes zero stiffness during slippage is shown in Fig. 2.14 in the form of an algorithm.

With reference to the system proposed by Pall and Marsh (1982), in which the braces in a moment resisting frame incorporated frictional devices, for example a typical X-braced system, the braces are designed to buckle at relatively low compressive loads. As a result, the braces contribute only when subjected to tension. By installing uniaxial friction elements within each brace, slippage would only occur in the tensile direction and very little energy dissipation would result during the cyclic loading. However, the special damper mechanism, depicted in

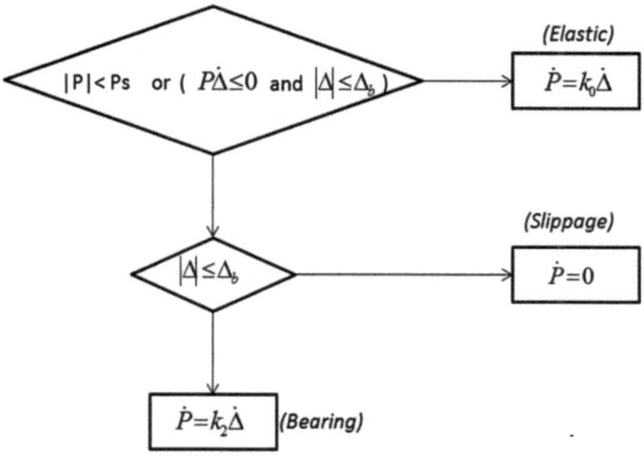

Fig. 2.14 Hysteretic model

Fig. 2.15, permits much more effective operation. During cyclic loading, the mechanism tends to straighten buckled braces and also enforces slippage in both tensile and compressive directions. Initially, Pall and Marsh (1982) used a simple elastoplastic model to represent the behavior of this X-braced friction damper. However, Filiatrault and Cherry (1987) determined that this is only valid if the device slips during every cycle, and if that the slippage is always sufficient to completely straighten any buckled braces. Otherwise, the Pall-Marsh model overestimates the energy dissipation. To remedy this situation, Filiatrault and Cherry (1989, 1990) proposed a more detailed model for the device. Each member of the bracing-damper system is represented by elements reflecting its individual axial and bending characteristics. Thus, the structural braces are assumed to yield

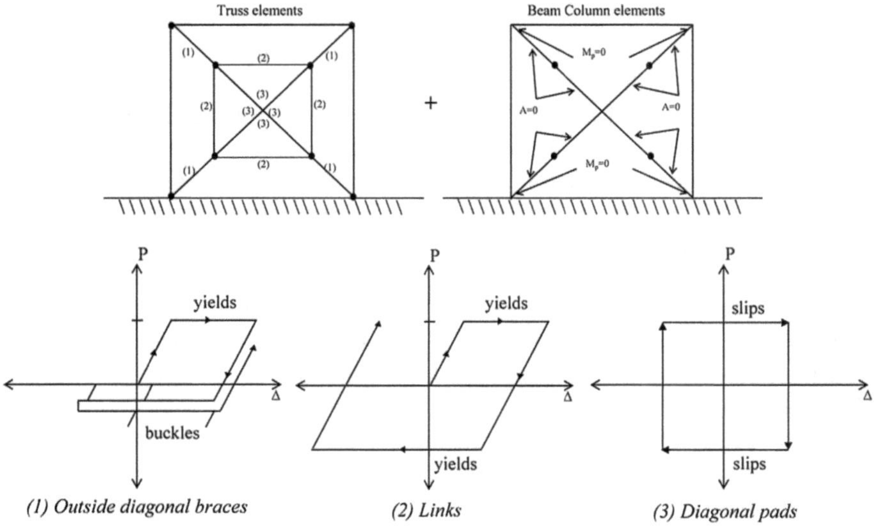

Fig. 2.15 Refined model for X-braced friction damper (modified from Soong and Dargush 1997)

in tension, but buckle elastically in compression. The device links (links 2) are permitted to yield in both tension and compression, while the sliding brake pads are represented by a hysteretic model corresponding to the experimental results obtained by Pall et al. (1980). The Filiatrault and Cherry (1989, 1990) and Pall and Marsh (1982) models are shown in Fig. 2.15.

Regarding the Slotted Bolted Connection (SBC), shown in Fig. 2.10e, this device was investigated by FitzGerald et al. (1989). Several experimental results obtained under sizeable displacement controlled loading have developed by FitzGerald et al. (1989) and Grigorian et al. (1993). The experimental results have shown that there can also be slippage between the channel and cover plates, which suolh be considered in the mathematical models.

As regards the uniaxial friction damper illustrated in Fig. 2.10c, manufactured by Sumitomo Metal Industries Ltd., it utilizes a slightly more sophisticated design. The pre-compressed internal spring exerts a force that is converted through the action of inner and outer wedges into a normal force on the friction pads. These copper alloy friction pads contain graphite plug inserts, which provides dry lubrication. This helps to maintain a consistent coefficient of friction between the pads and the inner surface of the steel casing. Aiken and Kelly (1990) indicate that the response of these dampers is extremely regular and repeatable with rectangular hysteresis loops. Furthermore, the effect of loading frequency and amplitude, number of cycles, or ambient temperature on damper response was reported to be insignificant. For structural analysis involving this device, a simple elastic-perfectly plastic hysteretic model, defined in Fig. 2.14, is appropriate.

Finally, with reference to the Energy Dissipating Restraint (EDR) manufactured by Fluor Daniel, Inc. and detailed in Fig. 2.10d, superficially, the design is similar to the Sumitomo concept, since this device also includes an internal spring and wedges encased in a steel cylinder. However, there are several novel aspects of the EDR that combine to produce very different response characteristics. As indicated in Fig. 2.10d, the EDR utilizes steel compression wedges and bronze friction wedges to transform the axial spring force into a normal pressure acting outward on the cylinder wall. Thus, the frictional surface is formed by the interface between the bronze wedges and steel cylinder. Internal stops are provided within the cylinder in order to create the tension and compression gaps. Consequently, unlike the Sumitomo device, the length of the internal spring can be altered during operation, providing a variable frictional slip force. Typical experimental hysteretic behavior have been developed for three different configurations to study: the response obtained with zero gaps and zero spring preload; triangular shaped hysteresis loops response with slip force proportional to the device displacement; with non-zero spring preload and very large gaps, the device acts as a standard Coulomb damper.

The model presented previously in Fig. 2.14 is obviously applicable for this second case. Finally, with a non-zero preload, but no initial gaps, the flag-shaped hysteresis loops are obtained.

Consider the case having zero gaps and zero preload. In its initial state, due to a zero internal spring force, there exists no normal contact pressure acting between the wedges and casing. However, once a force P is applied in either tension or compression, the spring, with stiffness K_s, is compressed and frictional resistance results. Let the spring displacement be represented by Δ_s, while the overall displacement of the device is Δ, which includes deformation of the rod and connections Δ_r. Thus (Soong and Dargush 1997),

$$\Delta = \Delta_s + \Delta_r \tag{2.17}$$

If the stiffness of the rod and connections equals K_3, then

$$P = K_3\Delta_r = K_1\Delta \tag{2.18}$$

where K_1 is the effective overall stiffness of the device during initial loading. Furthermore, the spring force becomes:

$$P_s = K_s\Delta_s \tag{2.19}$$

and the corresponding frictional resistance during slippage can be written:

$$P_f = \alpha K_s\Delta_s \tag{2.20}$$

The positive factor α, which is less than one for practical designs, incorporates the geometric and Coulomb friction effects involved in transforming the action of the spring force through the wedges into a frictional resistance. For slippage during loading, equilibrium requires that:

$$P = P_S + P_f \tag{2.21}$$

Consequently, from Eqs. (2.17)–(2.21), it is possible to obtain the following expression for the effective stiffness (Soong and Dargush 1997):

$$K_1 = \frac{(1+\alpha)K_sK_3}{(1+\alpha)K_s + K_3} \tag{2.22}$$

This is simply the stiffness of a system featuring a parallel combination of internal spring and frictional elements, $(1+\alpha)K_s$, in series with the rod/connection spring K_3. Upon subsequent unloading of the device, the frictional force reduces immediately and further slippage is prevented. Thus, for the initial stage of unloading, the spring displacement remains constant. The stiffness of the device is then simply equal to the stiffness of the rod and connections K_3. As the applied force P is reduced, a level is reached at which slippage occurs in the unloading direction. In this regime, the frictional force now opposes the action of the internal spring. The effective stiffness of the device becomes (Soong and Dargush 1997):

$$K_2 = \frac{(1-\alpha)K_sK_3}{(1-\alpha)K_s + K_3} \tag{2.23}$$

Note that in this configuration, the EDR device is self-centering. In the absence of external force, the internal spring will return to its initial zero preload state.

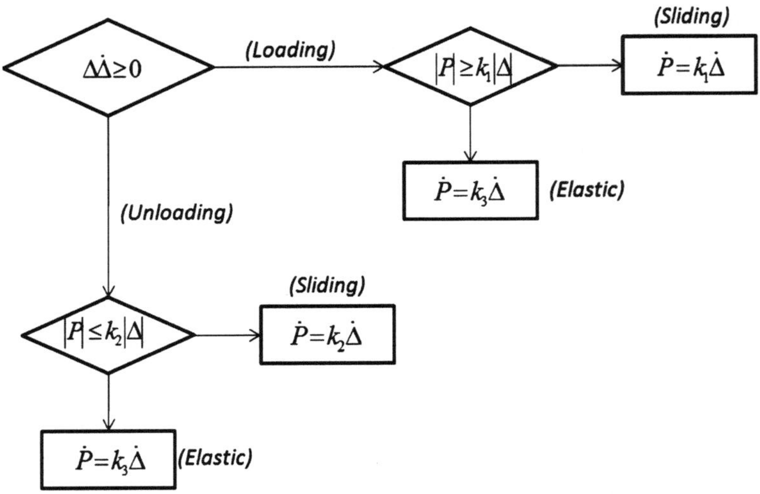

Fig. 2.16 Hysteretic model for EDR with no gaps and no prelaod

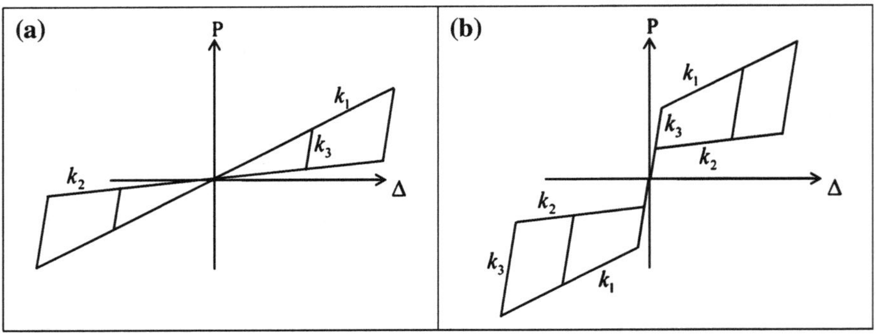

Fig. 2.17 Hysteretic model for EDR: **a** No gaps, no prelaod; **b** No gaps, finite preload (modified from Soong and Dargush 1997)

A detailed description of an appropriate hysteretic model is presented in Fig. 2.16. The results obtained from the model for displacement controlled cyclic loading are shown in Fig. 2.17.

2.4 Viscoelastic Dampers

The application of viscoelastic materials to vibration control can be dated back to the 1950s when it was first used on aircraft as a means of controlling vibration-induced fatigue in airframes (Ross et al. 1959). Since that time, it has been widely used in aircrafts and aerospace structures for vibration reduction. Its application to civil engineering structures appears to have begun in 1969 when 10,000 viscoelastic dampers were installed in each of the twin towers of the World Trade Center in New York to help resist wind loads. Seismic applications of viscoelastic dampers have a more recent origin. For seismic applications, larger damping increases are usually required in comparison with those required for mitigation of wind-induced vibrations. Furthermore, energy input into the structure is usually spread over a wider frequency range, requiring more effective use of the viscoelastic materials.

Viscoelastic materials used in structural application are typically copolymers or glassy substances which dissipate energy when subjected to shear deformation. A typical viscoelastic (VE) damper is shown in Fig. 2.18 which consists of viscoelastic layers bonded with steel plates.

If mounted in a structure, shear deformation and hence energy dissipation takes place when the structural vibration induces relative motion between the outer steel flanges and the center plate.

For these dampers, the shear stress can also be written as:

$$\tau(t) = \gamma_0[G'(\omega)\sin\omega t + G''(\omega)\cos\omega t] \tag{2.24}$$

where

$$G'(\omega) = \frac{\tau_0}{\gamma_0}\cos\delta \tag{2.25}$$

$$G''(\omega) = \frac{\tau_0}{\gamma_0}\sin\delta \tag{2.26}$$

where, as shown in Fig. 2.19, γ_0 and τ_0 are, respectively, the peak shear strain and peak shear stress, and δ is the phase angle.

Figure 2.19 shows the response of a viscoelastic damper, which defines an ellipse, whose area gives the energy dissipated by the viscoelastic material per unit volume and per cycle of oscillation. It is given by:

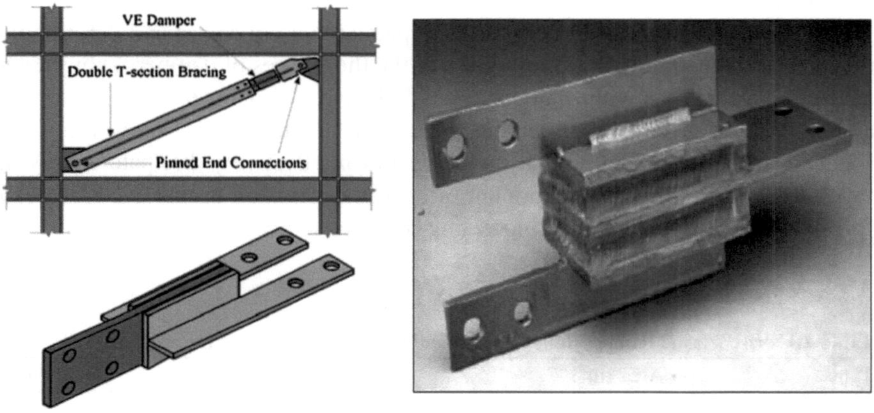

Fig. 2.18 Typical damper configuration

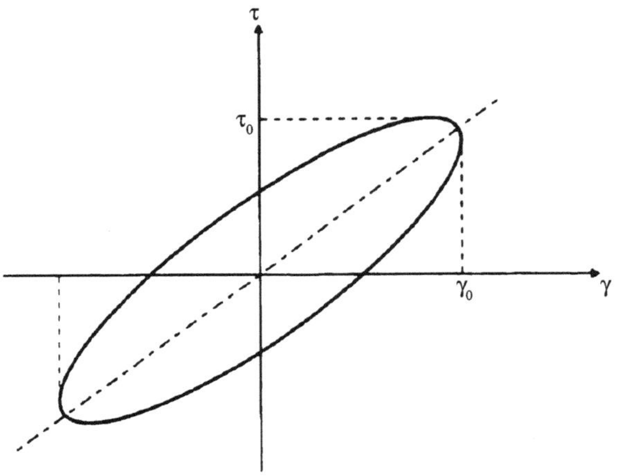

Fig. 2.19 Plot of stress versus strain

$$E_D = \int_0^{\frac{2\pi}{\omega}} \tau(t)\dot{\gamma}(t)dt = \pi\gamma_0^2 G''(\omega) \tag{2.27}$$

It is seen that the two moduli, G′ and G‴ defined respectively shear storage modulus and shear loss modulus, determine the dynamic behavior of the linear viscoelastic material in shear under time harmonic excitation. These moduli are not only functions of the excitation frequency (ω), but also functions of the ambient temperature (T) and, sometimes, the shear strain (γ) as well as the internal temperature (θ). One of the expedient ways of estimating their dependence on these parameters is to perform experiments on viscoelastic specimens over representative ranges of these variables. In one series of these tests (Chang et al. 1993), three types of viscoelastic dampers with configurations (A, B and C) distinguished by dimensions and types of the viscoelastic material were used. Type A and B dampers are made of similar VE materials but different in damper dimensions. The type C damper is made from a different VE material. In Table 2.1, the area, thickness and volume of each type of the dampers are reported.

The devices were tested at different temperatures, frequencies and deformation range. The force–deformation responses of the three types of dampers subjected to sinusoidal excitations with frequency of 3.5 Hz and 5 % damper strain at different ambient temperatures have demonstrated that the hysteresis loops are fairly rounded in shape, indicating that the dampers can effectively dissipate energy. In fact, the damper stiffness and amount of energy dissipation in one cycle decrease for all types of dampers with increasing ambient temperature. Thus, both shear

Table 2.1 Viscoelastic damper dimensions (modified from Soong and Dargush 1997)

Type	Area (m^2)	Thickness (m)	Volume (m^3)
A	0.00097	0.005	4.9×10^{-6}
B	0.0019	0.007	1.5×10^{-5}
C	0.0116	0.003	4.4×10^{-5}

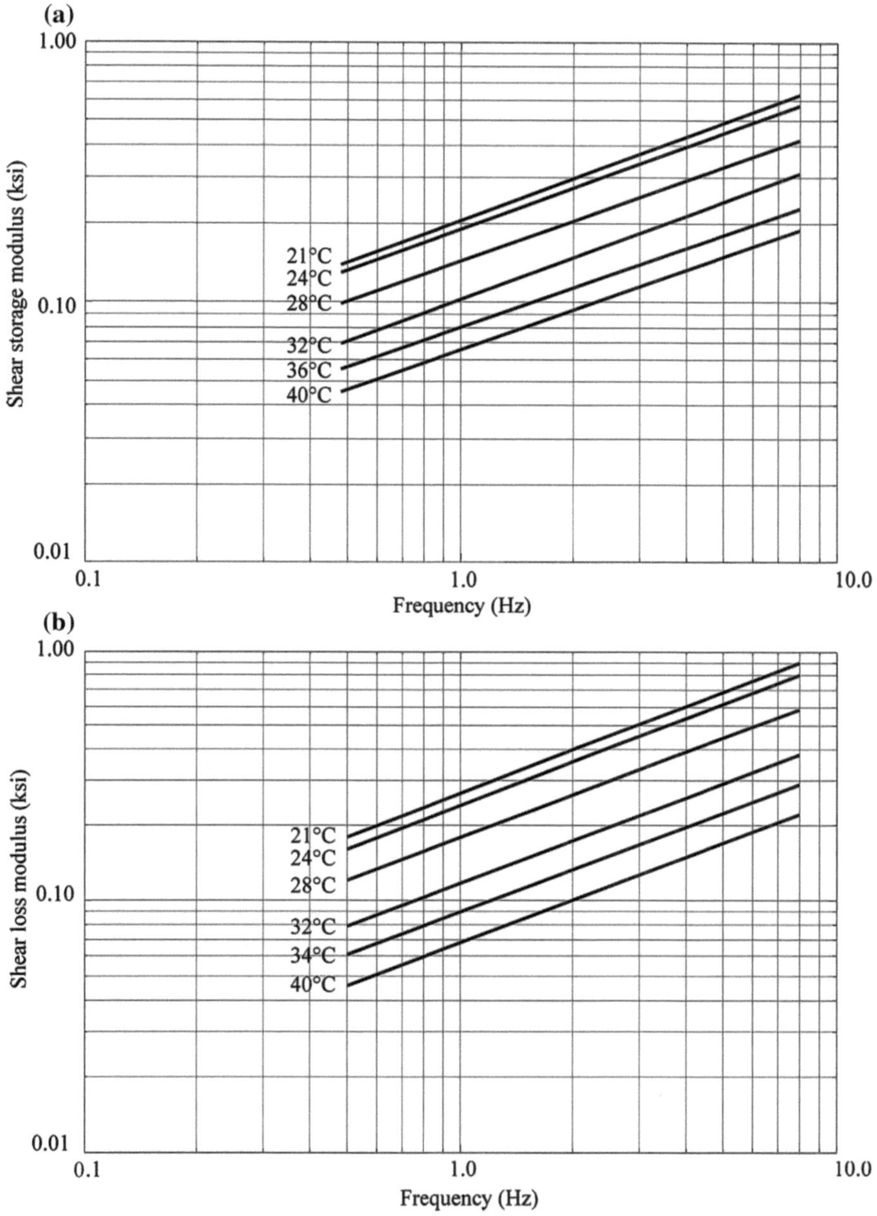

Fig. 2.20 Dependence of shear storage modulus **a** and shear loss modulus **b** on frequency and temperature (modified from Soong and Dargush 1997)

storage modulus and shear loss modulus decrease as the temperature increases, while they increase with increasing frequency as shown in Fig. 2.20a and b.

Field observations and laboratory experiments have shown that, during each wind and earthquake loading cycle, the transient temperature increase is typically less than 10 °C and has a minor effect on the performance of VE dampers.

From Fig. 2.20, it is noted that for a given temperature and moderate deformations, the stress in a viscoelastic material is linearly related to the deformation and strain rate when subjected to a harmonic motion. Experimental tests (Bergman and Hanson 1993; Lobo et al. 1993; Chang et al. 1995) have shown that the viscoelastic behavior of the damper can be modeled using the Kelvin model of viscoelasticity.

2.5 Viscous Fluid Dampers

The VF damper, widely used in the military and aerospace industry for many years, has recently been adapted for structural applications in civil engineering. In fact, a major reason for the relatively rapid pace of implementation of viscous fluid dampers is their long history of successful application in the military. Shortly after the Cold War ended in 1990, the technology behind the type of fluid damper that is most commonly used today (i.e., dampers with fluidic control orifices) was declassified and made available for civilian use (Lee and Taylor 2001). Applying the well-developed fluid damping technology to civil structures was relatively straightforward to the extent that, within a short time after the first research projects were completed on the application of fluid dampers to a steelframed building (Constantinou and Symans 1993) and an isolated bridge structure (Tsopelas et al. 1994).

A VF damper generally consists of a piston within a damper housing filled with a compound of silicone or similar type of oil, with the piston containing a number of small orifices through which the fluid may pass from one side of the piston to the other (Constantinou and Symans 1993). As the damper piston rod and piston head are stroked, fluid is forced to flow through orifices either around or through the piston head.

The resulting differential in pressure across the piston head (very high pressure on the upstream side and very low pressure on the downstream side) can produce very large forces that resist the relative motion of the damper (Lee and Taylor 2001). The fluid flows at high velocities, resulting in the development of friction between fluid particles and the piston head. The friction forces give rise to energy dissipation in the form of heat. Thus, VF dampers dissipate energy through the movement of a piston in a highly VF based on the concept of fluid orificing. Figure 2.21 contains a typical longitudinal cross sections of viscous fluid dampers.

Regarding a pure viscous behavior, the damper force and velocity should remain in phase. However, for a damper setup shown in Fig. 2.21, the volume for storing the fluid will change while the piston begins to move. Thus a restoring

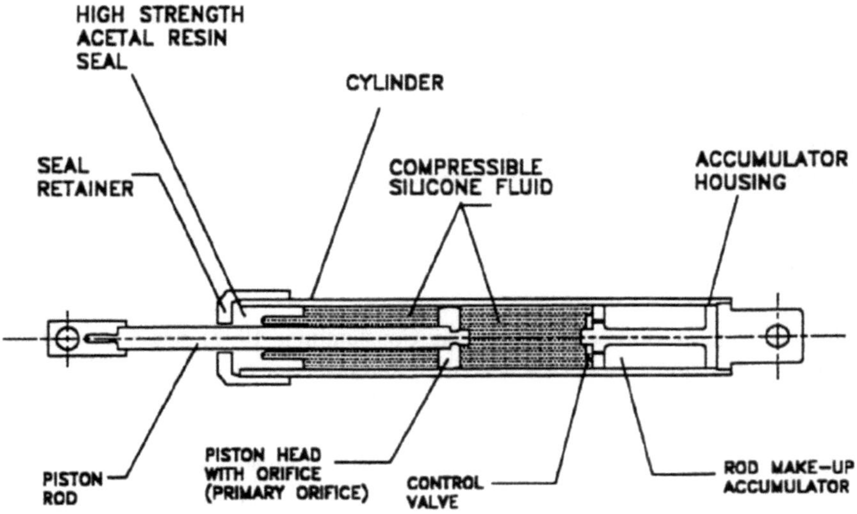

Fig. 2.21 Longitudinal cross section of a fluid damper

force, which is in phase with displacement rather than velocity, will be developed. Configuration of an accumulator or a run-through rod is used to solve the problem. However, for high frequency motions, the accumulator valve may operate inaccurately, and the restoring force will occur.

Additionally, for these devices, the associated temperature increase can be significant, particularly when the damper is subjected to long-duration or large-amplitude motions (Makris 1998). Mechanisms are available to compensate for the temperature rise such that the influence on the damper behavior is relatively minor (Soong and Dargush 1997). However, the increase in temperature may be of concern due to the potential for heat-induced damage to the damper seals. In this case, the temperature rise can be reduced by reducing the pressure differential across the piston head (e.g., by employing a damper with a larger piston head) (Makris et al. 1998). Interestingly, although the damper is called a viscous fluid damper, the fluid typically has a relatively low viscosity (e.g., silicone oil with a kinematic viscosity on the order of 0.001 m^2/s at 20 °C). The term viscous fluid damper is associated with the macroscopic behavior of the damper which is

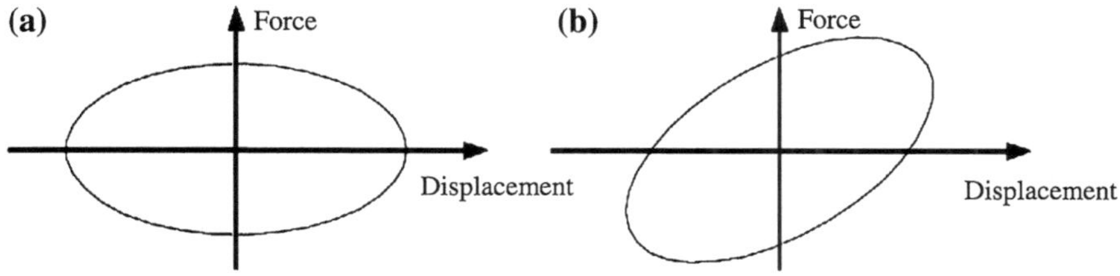

Fig. 2.22 Hysteresis loops of dampers with pure viscous and viscoelastic behavior

essentially the same as that of an ideal linear or non-linear viscous dashpot (i.e., the resisting force is directly related to the velocity).

Figure 2.22a shows the hysteresis loop of a pure linear viscous behavior. The loop is a perfect ellipse. The absence of storage stiffness makes the natural frequency of a structure incorporated with the damper remain the same. This advantage will simplify the design procedure for a structure with supplemental viscous devices.

However, if the damper develops restoring force, the loop will be changed from Fig. 2.22a, b. In other words, it turns from a viscous behavior to a viscoelastic behavior.

Experimental testing (Seleemah and Constantinou 1997) has shown that a suitable mathematical model for describing the behavior of viscous fluid dampers is given by the following non-linear force–velocity relation:

$$F_d(t) = c \cdot (\dot{u}_d(t))^{\alpha} = c \cdot |\dot{u}_d(t)|^{\alpha} \mathrm{sgn}[\dot{u}_d(t)] \tag{2.28}$$

where $F_d(t)$ is force developed by the damper; $u_d(t)$ is displacement across the damper; c is damping coefficient; α is exponent whose value is determined by the piston head orifice design; sgn [·] is signum function, and the overdot indicates ordinary differentiation with respect to time, t. The physical model corresponding to Eq. (2.28) is a non-linear viscous dashpot. For earthquake protection applications, the exponent typically has a value ranging from about 0.3 to 1.0. For α equal to unity, the damper may be described as an ideal linear viscous dashpot. The damper with $\alpha = 1$ is called a linear viscous damper in which the damper force is proportional to the relative velocity. The dampers with α larger than 1 have not often been seen in practical applications. The damper with α smaller than 1 is called a non-linear viscous damper which is effective in minimizing high velocity shocks.

Figure 2.23 shows the force–velocity relationships of the three different types of viscous dampers. This Figure demonstrates the efficiency of non-linear dampers in minimizing high velocity shocks.

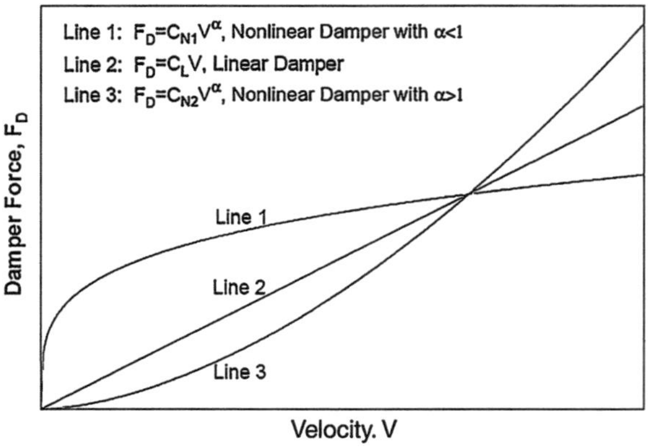

Fig. 2.23 Force–velocity relationships of viscous dampers

For a small relative velocity, the damper with a α value less than 1 can give a larger damping force than the other two types of dampers.

The energy dissipated per cycle of steady-state harmonic motion, characterized by a frequency ω, is obtained by integrating Eq. (2.28) over the displacement leading to the following expression (Symans and Constantinou 1998):

$$E_D = 4F_{d,0}u_{d,0}2^{\alpha}\left(\frac{\Gamma^2(1+\alpha/2)}{\Gamma(2+\alpha)}\right) = \lambda F_{d,0}u_{d,0} \tag{2.29}$$

where $F_{d,0}$ = peak force developed by the damper ; $u_{d,0}$ = peak displacement across the damper; Γ = gamma function; λ = parameter whose value depends exclusively on the velocity exponent. For a given force and displacement amplitude, the energy dissipated per cycle for a non-linear fluid damper is larger, by a factor λ/π, than that for the linear case and increases monotonically with reducing velocity exponent (up to a theoretical limit of $4/\pi = 1.27$ which corresponds to a velocity exponent of zero) (Fig. 2.24). For a given frequency of motion and displacement amplitude, $u_{d,0}$, to dissipate the same amount of energy per cycle, the damping coefficient of the non-linear damper, c_{NL}, must be larger than that of the linear damper, c_L, as given by (Symans et al. 2008):

$$c_{NL} = c_L\frac{\pi}{\lambda}\left(\omega u_{d,0}\right)^{1-\alpha} \tag{2.30}$$

As an example, for a frequency of 1.0 Hz and displacement amplitude of 5 cm (approximately 2 % story drift if the dampers are installed horizontally within a chevron brace configuration), the damping coefficient of a non-linear damper with velocity exponent of 0.5 must be approximately three times larger than that of a linear damper to dissipate the same amount of energy per cycle (Symans et al. 2008).

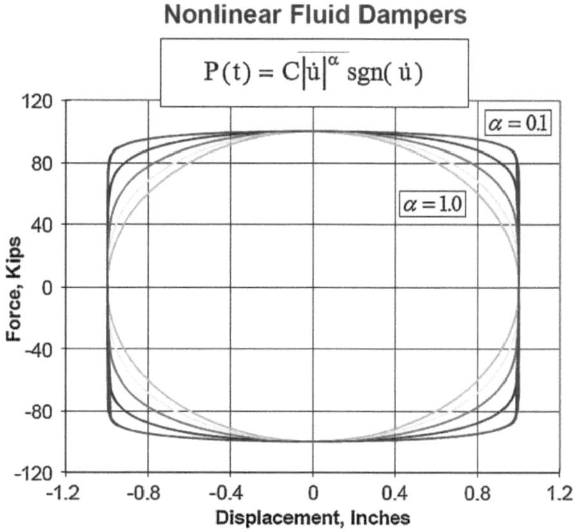

Fig. 2.24 Force–displacement relationships of viscous linear and non linear dampers

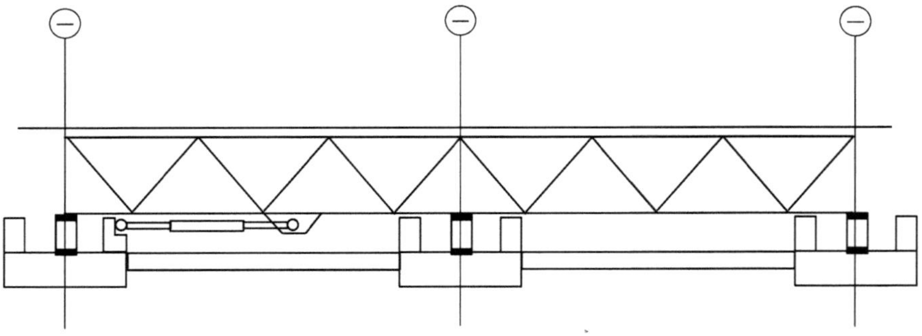

Fig. 2.25 San Bernardino County Medical Center–damper–base isolation system assembly (modified from Symans et al. 2008)

Conversely, if non-linear dampers are used to limit the damper force and thus the base shear, a reduction in the energy dissipation capacity as compared to the case of linear dampers would be accepted to ensure that the base shear is limited.

Note that an expression equivalent to Eq. (2.30) has been derived by Filiatrault et al. (2001), wherein it is explained that, having identified suitable linear damping coefficients to meet some design criterion, the equation can be used to estimate initial values of non-linear damping coefficients.

The viscous fluid (VF) devices developed recently include viscous walls and VF dampers (Soong and Spencer 2002). The viscous wall, developed by the Sumitomo Construction Company, consists of a plate moving in a thin steel case filled with a high VF.

In several applications, they have been used in combination with seismic isolation systems to prevent the system from a large deformation (Symans et al. 2008; Hwang and Huang 2003). For example, in 1995, VF dampers were incorporated into base isolation systems for five buildings of the San Bernardino County Medical Center, located close to two major fault lines. The five buildings required

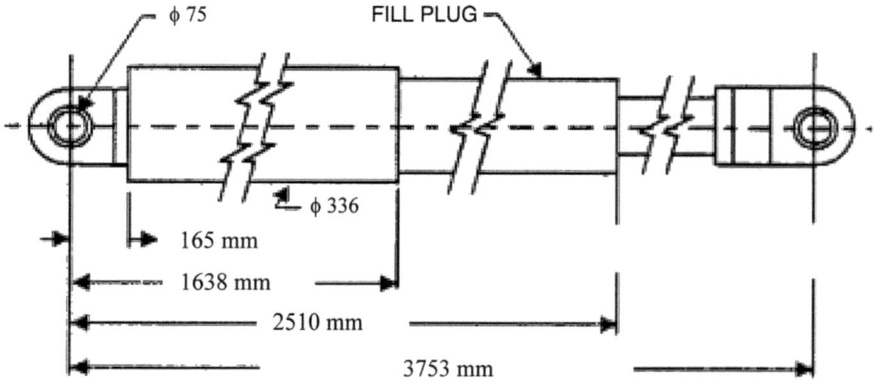

Fig. 2.26 Dimensions of VF damper for San Bernardino County Medical Center (modified from Symans et al. 2008)

a total of 233 dampers, each having an output force capacity of 320,000 lb and generating an energy dissipation level of 2,206 kw at a speed of 60 in/s. A layout of the damperisolation system assembly is shown in Figs. 2.25 and 2.26 gives the dimensions of the viscous dampers employed.

When the viscous fluid dampers are included in a strucutal systems, their response to lateral loads improves by increasing the damping factor from 2–5 %, typical of constructions, to 20–30 % of the critical value, this greatly reduces the accelerations and displacements of the structure.

Finally, as an alternative to viscous fluid dampers, viscoelastic fluid dampers, which are intentionally designed to provide stiffness in addition to damping, have recently become available for structural applications (Miyamoto et al. 2003). These dampers provide damping forces via fluid orificing and restoring forces via compression of an elastomer. Thus, more accurately, the dampers may be referred to as viscoelastic fluid/solid dampers.

2.5.1 Design Considerations for Viscous Passive Energy Dissipation Systems

Retrofit applications of passive damping systems have been used to limit the inelastic demands of connections in both steel and concrete moment frames (Symans et al. 2008). Providing retrofit improvements in this manner can be very cost competitive when compared to a conventional approach of retrofitting each welded connection to improve deformation capacity. By assuming a perfectly rigid bracing-damper system and associated connections and considering the elastic structural response, the linear viscous dampers produce forces within a given story that are 90° out of phase with respect to the restoring forces in the same story. Therefore, the viscous force is maximum when other forces are zero. This is a feature that can be exploited in the case of existing structures which might not have enough capacity to carry out the exerted forces. In this case, for retrofit applications in which the damping is proportionally distributed, and considering only the response in the fundamental mode, the impact of the damping forces on the existing foundation may be minor and therefore the foundation, which is usually very difficult and expensive to retrofit, may require minimal, if any, strengthening. A staggered distribution of devices in elevation may increase the axial force in the columns and thereby reduce their resistance capacity to absorb the bending stresses making them more vulnerable. A more rational distribution based on a continuity in the transmission of the effort up to the foundations can eliminate this effect.

In reality, elastic structure forces and viscous damping forces are usually partially in phase, leading to the possibility of increased forces at the foundation level (Constantinou et al. 1998; Fu and Kasai 1998). The partially in-phase relation for the elastic and viscous damping forces can be induced by damper bracing and connection flexibility (Constantinou et al. 1998; Fu and Kasai 1998), higher mode effects, and nonproportional damping effects.

It is also important to recognize that, for strong earthquakes, most structures employing viscous dampers will experience some level of inelastic response in the structure framing system. In this case, damping forces and inelastic restoring forces may be additive, causing significant increases in the base shear. Adding dampers to a structure introduces a new and very important design requirement in that the deformations along the load path between all the dampers and main structural elements must be included in the analysis (e.g., a rigid diaphragm action cannot be assumed). Failure to account for such deformations can reduce the effectiveness of the damping system to the point where the damping system simply rides along with the seismic movements and provides virtually no response reduction (Fu and Kasai 1998; Lin and Chopra 2003; Charney and McNamara 2002, 2008).

2.5.1.1 The Loading Combination Factors CF_1 and CF_2 according to FEMA 273

For a linear elastic simple-degree-of-freedom structure with a (linear or non linear) damper under a harmonic vibration at its natural frequency, ω, the displacement and velocity may be expressed as (Hwang and Huang 2003; Hwang et al. 2008; Hwang 2005):

$$\begin{aligned} u &= u_0 \cos \omega t \\ \dot{u} &= -\omega u_0 \sin \omega t \end{aligned} \tag{2.31}$$

The force response should be:

$$F = ku + c|\dot{u}|^{\alpha} \operatorname{sgn}(\dot{u}) \tag{2.32}$$

where k is the stiffness of the system and is equal to $m\omega^2$. Figure 2.27 shows this force–displacement relationship.

In the case of linear devices, to describe the viscoelastic overall response, it is possible to write:

$$F = F_0 \cos(\omega t + \delta) \tag{2.33}$$

where F is the resistance force of the system; F_0 is amplitude of the force; and the δ is the phase angle.

Since the force of viscous dampers and the displacement response of the frame are out of phase, it is difficult to determine the internal force of each member of the frame through the static procedure. Therefore, when the rehabilitation of buildings is executed with velocity dependent devices, FEMA 273 (ATC 1997a) suggests that engineers check the actions for components of the buildings in the following three stages of deformation, and the maximum action should be used for design.

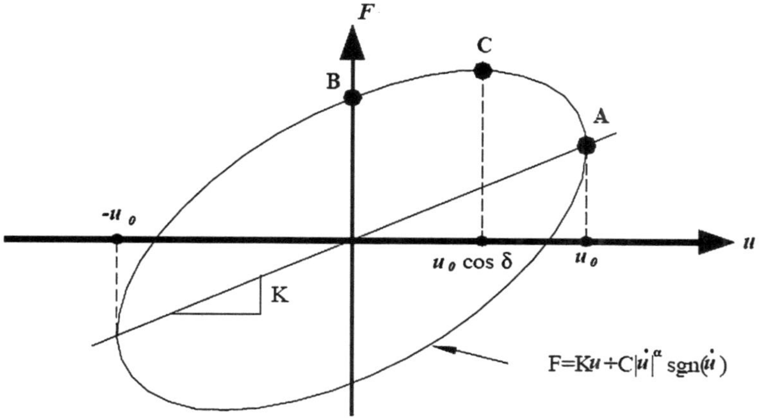

Fig. 2.27 Force–displacement relationship of a structure with viscous dampers

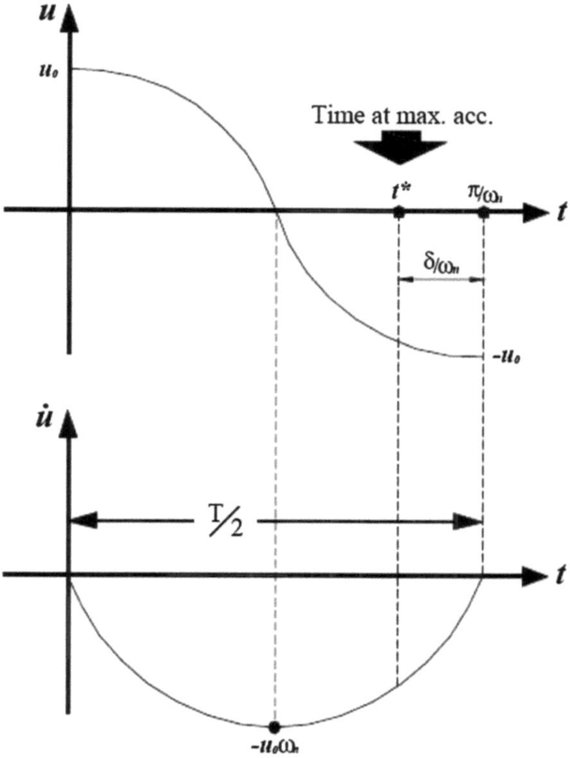

Fig. 2.28 Harmonic motion of a structure with viscous dampers

- Stage of maximum drift: which is represented by point A of Fig. 2.27.
- Stage of maximum velocity and zero drift: which is represented by point B of Fig. 2.27.
- Stage of maximum floor acceleration: which is represented by point C of Fig. 2.27.

Figure 2.28 shows an example of response in terms of the two magnitudes, displacement and velocity, and the instants in which there is the maximum of the displacement, velocity and acceleration respectively.

Furthermore, FEMA 273 (ATC 1997a) recommends a procedure to calculate the member force at the instant of the maximum acceleration. The procedure indicates that design actions in components should be determined as the sum of "actions determined at the stage of maximum drift" times CF_1 and "actions determined at the stage of maximum velocity" times CF_2, where:

$$\begin{aligned} CF_1 &= \cos\left(\tan^{-1}(2\xi)\right) \\ CF_2 &= \sin\left(\tan^{-1}(2\xi)\right) \end{aligned} \tag{2.34}$$

where ξ is the damping of the overall system, equal to the sum of the internal damping ξ_s of the structure and the damping produced by dampers ξ_d:

$$\xi = \xi_s + \xi_d \tag{2.35}$$

For the case of linear dampers ($\alpha = 1$), the following relation can be written:

$$\delta = \tan^{-1}(2\xi_d) \tag{2.36}$$

which is adopted by FEMA 273 (ATC 1997a) for linear dampers.

However, the two load combination factors are inappropriate for structures with non-linear viscous dampers. The revised formulas have been proposed by Ramirez et al. (2001) and are implemented in NEHRP 2000 (BSSC 2004). The following are the derivation of the revised load factors:

$$\begin{aligned} CF_1 &= \cos\delta \\ CF_2 &= \sin^{\alpha}\delta \end{aligned} \tag{2.37}$$

2.6 Comparison Between Passive Energy Dissipation Devices

A summary of passive energy dissipation devices that have been commonly used is presented in Fig. 2.29, with the construction, hysteretic behavior, physical models, advantages, and disadvantages being presented.

The fiction devices are characterized by large energy dissipation per cycle and insensitivity to ambient temperature, but have a strongly non-linear behavior and require non-linear analysis. Sliding interface conditions may change with time

	Viscous damper	Viscoelastic damper	Metallic damper	Friction damper
Basic Conctruction		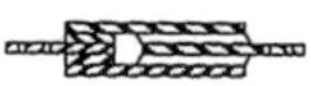	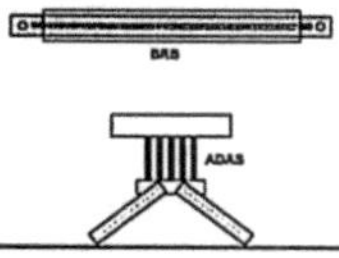	
Idealized Hysteretic Behavior	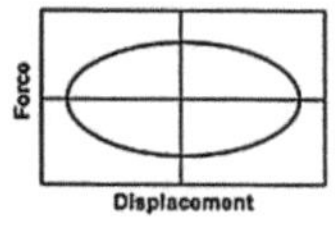	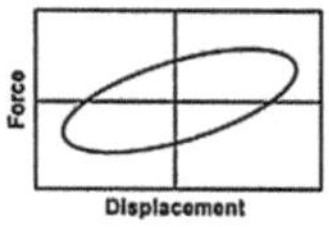	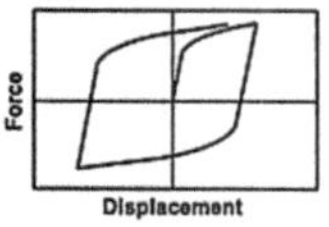	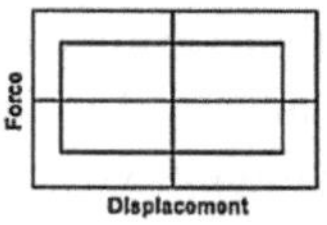
Idealized Physical Model	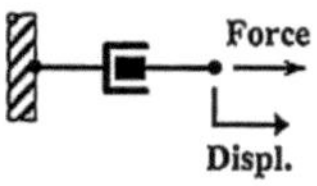	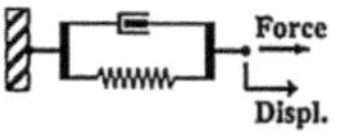	Idealized Model Not Available	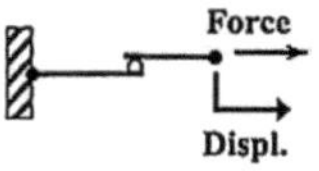
Advantages	• Activated at low displacements • Minimal restoring force • For linear damper, modeling of damper is simplified • Properties largely frequency and temperature-independent • Proven record of performance in military applications	• Activated at low displacements • Provides restoring force • Linear behavior, therefore modeling of damper is simplified	• Stable hysteretic behavior • Long-term reliability • Insensitivity to ambient temperature • Materials and behavior familiar to practicing engineers	• Large energy dissipation per cycle • Insensitivity to ambient temperature
Disadvantages	• Possible fluid seal leakage (reliability concern)	• Limited deformation capacity • Properties are frequency and temperature-dependent • Possible debonding and tearing of VE materil (reliability concern)	• Device damaged after earthquake; may require replacement • Non-linear behaviour; may require non-linear analysis	• Sliding interface conditions may change with time (reliability concern) • Strongly non-linear bahavior; may excite higher modes and require non-linear analysis • Permanent displacements if no restoring force mechanism provided

Fig. 2.29 Summary of construction, hysteretic behavior, physical models, advantages, and disadvantages of passive energy dissipation devices for seismic protection applications (modified from Symans et al. 2008)

and present permanent dispacements if no restoring force mechanism are provided. The metallic devices have also a non-linear behavior and are characterized by stable hysteresis behavior. They are not sensitive to ambient temperature and may require replacement due to damage in the post-earthquake phase. In relation to the viscoelastic devices, they provide restoring force and have a linear behavior, but their properties are frequency and temperature dependent. They have also a limited deformation capacity. Among the advantages of the viscous devices, it is worth noting activation at low displacements, they possess a restoring force and their mechanical properties are not frequency and temperature dependent. For these linear devices, a simplified modeling can be considered, in a first phase, with the assumption of not considering the flexibility of the bracing system and connections.

To demonstrate the different effectiveness of above-mentioned devices, a scientific research is reported (Tehrani and Maalek 2006), in which different strengthening methods are employed and compared through non-linear dynamic analysis. The investigated existing structure in this research is a nine story steel building with a rectangular plan and located in Tehran (Iran). The longest side is defined as Y-direction and the other one X-direction. More details on geometry and structural properties of the existing structure can be found in Tehrani and Maalek (2006).

The existing structure has been modeled in three dimensions in Perform-3D software and several non-linear static analyses have been performed on the model.

Regarding the results obtained on the existing structure, it has been possible to observe that the shear wall in the Y direction is failed by concrete crushing. The remarkable strength loss has been seen in the capacity curve that causes increase in the target displacement of the structure.

Additionally, in most of the connections, plastic deformations have exceed from the acceptable limits and the link beams in the eccentric bracings do not have enough capacity while their plastic rotation violates the limitation given in the FEMA356 (ATC 2000) instruction manual; therefore, in accordance with the FEMA356 guidelines (ATC 2000), the structure is vulnerable and needs to be rehabilitated.

Two groups of strengthening methods have been considered: the first group is related to the relatively modern methods based on the use of passive control devices that include metallic dampers, friction dampers, viscous dampers and

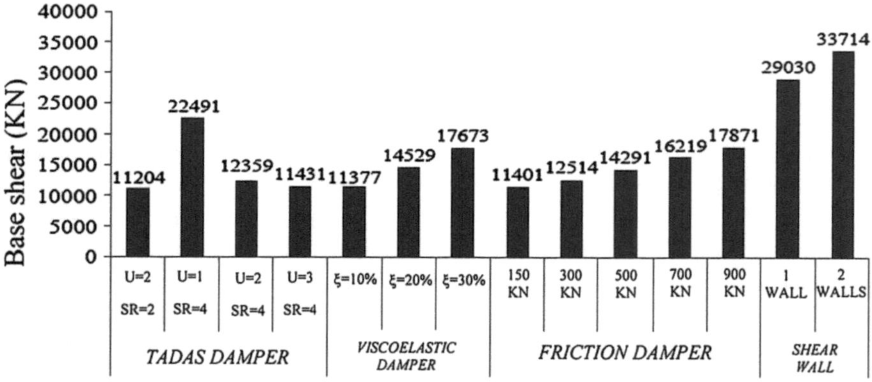

Fig. 2.30 Comparison of the maximum base shears for the use of various devices (X direction) (modified from Tehrani and Maalek 2006)

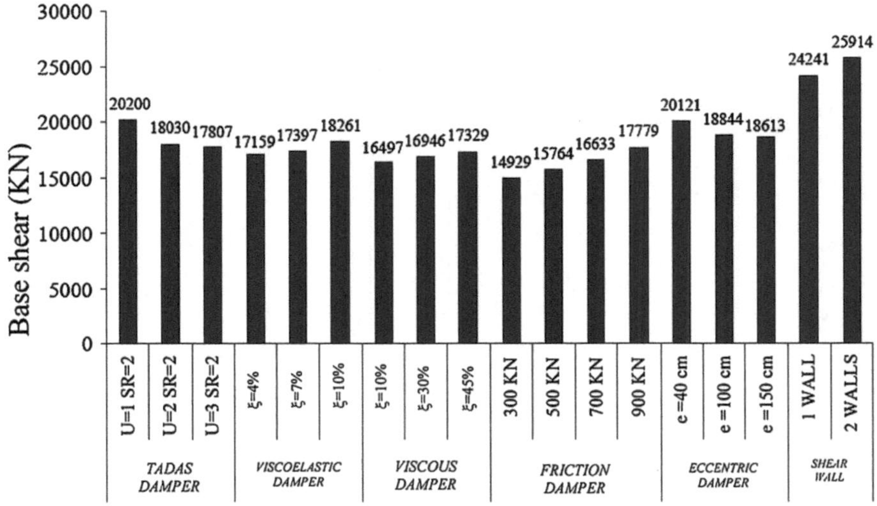

Fig. 2.31 Comparison of the maximum base shears for the use of various devices (Y direction) (modified from Tehrani and Maalek 2006)

viscoelastic dampers. While, on the other hand, conventional strengthening methods such as the use of bracings and shear walls have also been investigated.

For each of the rehabilitation schemes of the structure, a realistic model has been prepared and several non-linear dynamic analyses carried out on the models. The non-linear dynamic analyses have been performed using seven scaled earthquake records matched to the spectrum under consideration. These records include Naghan (Iran 1977), Tabas (Iran 1978), Abhar (Iran 1991), Elcentro (1940), Park field (1966), Taft (Kern County 1952), and San Fernando (1971) earthquakes.

In Figs. 2.30 and 2.31, the base shears of the structure for the use of various devices are presented.

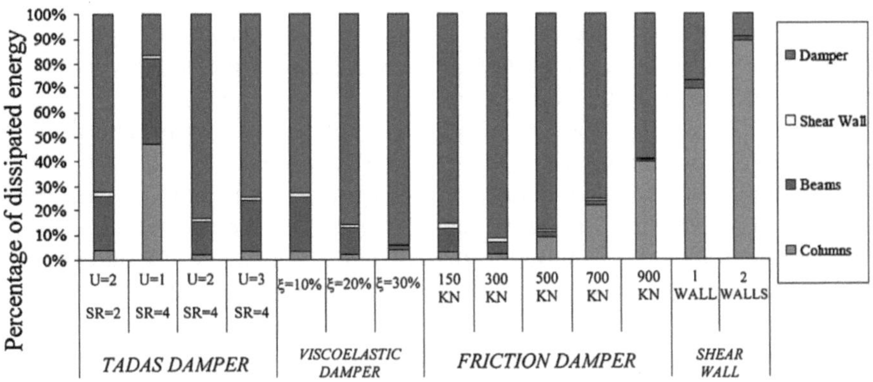

Fig. 2.32 Average dissipated energy in each element for the use of various devices (X direction) (modified from Tehrani and Maalek 2006)

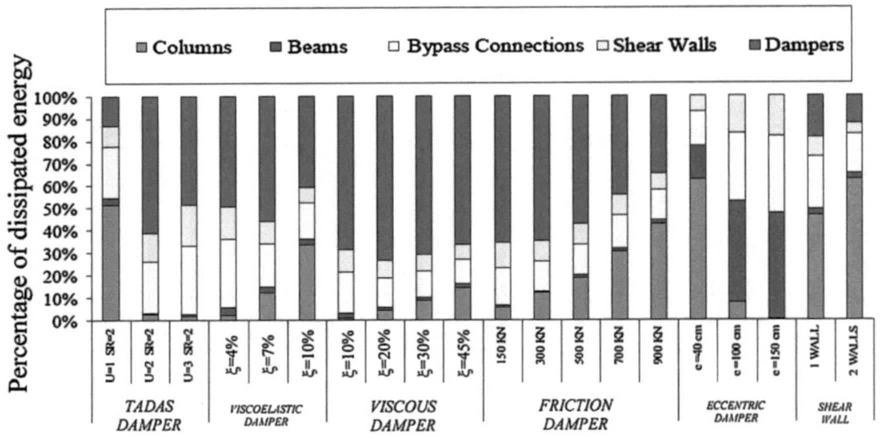

Fig. 2.33 Average dissipated energy in each element for the use of various devices (Y direction) (modified from Tehrani and Maalek 2006)

In the analysis, different values of the characteristics of TADAS dampers, which are SR and U values, defined as the relative strength and stiffness of the metallic TADAS dampers to the original frame of the structure respectively, have been considered. As for viscous and viscoelastic devices, the structure has been subjected to three different percentages of critical damping ξ for each direction separately. Therefore, to determine the optimum slipping load of the friction dampers, the structure has been analysed for the different values of slip loads. Then, as for conventional rehabilitation method, in the two perpendicular directions of the structure, the analyses have been undertaken with the consideration of two alternatives: the use of one and also two shear walls in the structure and by considering three values of the eccentricity bracing only in the Y direction.

By the comparison of these diagrams, it is possible to discover that the variation of the base shear forces is much higher in the X direction than in the Y direction. This is due to the existence of the stiff shear walls in this direction that decrease the effect of dampers to reduce the base shear of the structure. However these diagrams show that with the application of passive dampers, the base shears will be reduced, particularly in the X direction.

Figures 2.32 and 2.33 show the average percentage of dissipated energy in the structural elements in response to the seven earthquake records. The results show that the amount of dissipated energy in the VE dampers with $\xi = 0.30$ is maximum in direction X, while the other structural elements remain almost elastic. Also in this case, the structural damages in the columns are minimal. In addition, this diagram shows that the use of friction damper with the slip load of 300 KN can lead to the same results.

In direction Y, the results show that the amount of dissipated energy in the viscous damper with $\xi = 0.20$ is maximum. In this case, the damage in the columns and saddle (bypass) connections are minimal. This is because of the fact that viscous dampers exert their maximum forces in out of phase with displacement. Thus, when the structure undergoes its maximum deformations, no force is exerted by the viscous dampers to the structure. This desirable characteristic of viscous dampers is especially suitable for the existing structures that may not have enough capacity to carry out the exerted forces.

2.7 Design Recommendations: Guidelines

Guidelines for the implementation of energy dissipation or damping devices in new buildings were first proposed by the Structural Engineers Association of Northern California (SEAOC 1999) to provide guidance to structural engineers, building officials, and regulators who had the task of implementing such devices in building frames (Whittaker et al. 1993; Symans et al. 2008). In the mid-1990s, the Federal Emergency Management Agency (FEMA) funded the development of guidelines for the seismic rehabilitation of buildings (Kircher 1999). Four new methods of seismic analysis and evaluation were presented in the NEHRP Guidelines for the Seismic Rehabilitation of Buildings (FEMA Reports 273 and 274 ATC 1997a, b):

(1) linear static procedure,
(2) linear dynamic procedure;
(3) non-linear static procedure;
(4) non-linear dynamic procedure.

With regard to structures incorporating passive energy dissipation devices, the basic principles to be followed included in Guidelines (ATC 1996, 1997a, b) are:

(1) spatial distribution of dampers (at each story and on each side of the building);
(2) redundancy of dampers (at least two dampers along the same line of action);
(3) for maximum considered earthquake, dampers, and their connections designed to avoid failure (i.e., not the weak link in the system);
(4) members that transmit damper forces to foundation designed to remain elastic.

In addition, the following hypotheses (NEHRP (Recommended Provisions), ASCE/SIE 7-05 Standard entitled "Minimum design loads for buildings and other structures" (ASCE 2005), the 2006 International Building Code (ICC 2006) and the Building Construction and Safety Code (NFPA 2006)) were assumed:

(1) the collapse mechanism for the building is a single-degree-of-freedom mechanism so that the drift distribution over the height of the building can be reasonably estimated using either the first mode shape or another profile such as an inverted triangle;
(2) the building is analyzed in each principal direction with one degree-of-freedom per floor level;
(3) the non-linear response of the building can be represented by an elastoplastic relationship;

(4) the yield strength of the building can be estimated by either simple plastic analysis or using the specified minimum seismic base shear and values of the response modification (R), the reserve strength of the framing system (W_0), and the deflection amplification (C_d) factors presented in the NEHRP Recommended Provisions.

Note that the above-mentioned procedures for the analysis and design of structures with damping systems were largely based on studies that do not consider the effects of near-field (close to the fault) seismic excitations. However, as demonstrated by Pavlou and Constantinou (2004), the simplified methods of analysis for single-degree-of-freedom systems yield predictions of peak response of structures with damping systems that are generally accurate or conservative for the case of near-field seismic excitation (with a correction factor required for predicting peak velocity).

An engineering use of passive devices implies that their size and placement must be determined so that the predetermined performance are achieved. A traditional procedure for determining the size and position of the energy dissipation devices begins with the selection of values on the basis of the experience of the engineer. Thus, either a dynamic or static analysis must be performed to verify the achievement of the performance objective. To this scope, ATC-40 (ATC 1996) and FEMA-273 (ATC 1997b), then developed in FEMA-356 (ATC 2000), provide non-linear static analysis procedures that incorporate energy dissipation devices. If the analysis result is not satisfactory, the structure is analyzed again after changing the quantity or location of the devices. This process of trial and error is repeated until the optimal number of devices to achieve the performance is finally reached. However, the general practice of carrying out a series of tests and processes of error can be a laborious task unless the engineer has extensive experience in seismic design using supplemental energy dissipation devices. Several contributions in the literature relate to the development of more or less simplified design procedures of structural systems equipped with passive dissipation devices have been proposed according to the performance based design with the use of the method of the capacity spectrum or by using the response spectrum in terms acceleration-displacement (ADRS) or through the use of the response spectrum in terms of the displacement and capacity curve of the structure (Sullivan et al. 2003; Kim et al. 2003; Kim and Seo 2004; Lin et al. 2003; Kim and Choi 2006; ATC-40 1996 ATC 1997a; Freeman 1998; SEAOC 1999).

References

Aiken, I.D., Kelly, J.M.: Earthquake simulator testing and analytical studies of two energy absorbing systems for multistory structures, Report No. UCB/EERC-90/03, University of California, Berkeley (1990)

Aiken, I.D., Nims, D., Whittaker, A., and Kelly, J.: Testing of passive energy dissipation systems. Earthquake Spectra, Vol. 9, No. 3, Earthquake Engineering Research Institute, California (1993)

Applied Technology Council (ATC) (1996) Seismic evaluation and retrofit of concrete buildings, ATC-40, Redwood City, California
Applied Technology Council (ATC) (1996) Seismic evaluation and retrofit of concrete buildings, ATC-40, Redwood City, California
Applied Technology Council (ATC) (1997a) NEHRP commentary on the guidelines for the seismic rehabilitation of buildings, Report No. FEMA-273, Prepared for the Building Seismic Safety Council (BSSC) by the Applied Technology Council (ATC), Federal Emergency Management Agency (FEMA), Washington, D.C

Applied Technology Council (ATC) (1997b) NEHRP commentary on the guidelines for the seismic rehabilitation of buildings, Report No. FEMA-274, Prepared for the Building Seismic Safety Council (BSSC) by the Applied Technology Council (ATC), Federal Emergency Management Agency (FEMA), Washington, D.C

Applied Technology Council (ATC) (2000) Prestandard and Commentary for the Seismic Rehabilitation of Buildings, Report No. FEMA-356, Prepared for the Building Seismic Safety Council (BSSC) by the Applied Technology Council (ATC), Federal Emergency Management Agency (FEMA), Washington, D.C

ASCE (2005) Minimum design loads for buildings and other structures, ASCE/SEI 7–05, Reston, VA

Bergman, D.M., Hanson, R.D.: Viscoelastic mechanical damping devices tested at real earthquake displacements. Earthq. Spectra **9**(3), 389–418 (1993)

Black, C.J., Makris, N., Aiken, I.: Component testing, seismic evaluation and characterization of buckling-restrained braces. J. Struct. Eng. **130**(6), 880–894 (2004)

Building Seismic Safety Council (BSSC) (2004) NEHRP recommended provisions for seismic regulations for new buildings and other structures, 2003 Ed., Report Nos. FEMA-450/1 and FEMA-450/2, Prepared by the Building Seismic Safety Council (BSSC) for the Federal Emergency Management Agency (FEMA), Washington, D.C

Butterworth, J.J., Clifton, G.C.C.: Performance of hierarchical friction dissipating joints in moment resisting steel frames. In: Proceedings of 12 WCEE. http://www.iitk.ac.in/nicee/wcee/article/0718.pdf (2000)

Chang, K.C., Lai, M.L., Soong, T.T., Hao, D.S., Yeh, Y.C.: Seismic behavior and design guidelines for steel frame structures with added viscoelastic dampers, NCEER 93–0009. Buffalo, NY (1993)

Chang, K.C., Soong, T.T., Oh, S.-T., Lai, M.L.: Seismic behavior of steel frame with added viscoelastic dampers. ASCE **121**(10), 1418–1426 (1995)

Charney, F.A., Mcnamara, R.J.: Use of damped mode shapes to assess the seismic performance of a 39-story building with viscous fluid dampers. In: Proceedings of 7th U.S. National Conference on Earthquake Engineering, EERI, Oakland, California (2002)

Charney, F.A., Mcnamara, R.J.: A comparison of methods for computing equivalent viscous damping ratios of structures with added viscous damping. J. Struct. Eng. **134**(1), 32–44 (2008)

Constantinou, M.C., Symans, M.D.: Experimental study of seismic response of building with supplemental fluid dampers. J. Struct. Des. Tall Buildings **2**(2), 93–132 (1993)

Constantinou, M.C., Soong, T.T., Dargush, G.F.: Passive energy dissipation systems for structural design and retrofit, Monograph n.1, Multidisciplinary Center for Earthquake Engineering Research, Buffalo, New York (1998)

Fierro, E.A., Perry, C.L.: San Francisco retrofit design using added damping and stiffness (ADAS) elements. In: Proceedings of ATC-17-1 Seminar on Seismic Isolation, Passive Energy Dissipation and Active Control, vol. 2, pp. 593–603. ATC, Redwood City, California (1993)

Filiatrault, A., Cherry, S.: Performance evaluation of friction damped braced frames under simulated Earthquake loads. Earthq. Spectra **3**(1), 57–78 (1987)

Filiatrault, A., Cherry, S.: Efficient numerical modelling for seismic design of friction damped braced steel plane frames. Canadian J. Civ. Engrg. **16**(3), 211–218 (1989)

Filiatrault, A., Cherry, S.: Seismic design spectra for friction-damped structures. J. Struct. Engrg. **116**(5), 1334–1355 (1990)

Filiatrault, A., Tremblay, R., Wanitkorkul, A.: Performance evaluation of passive damping systems for the seismic retrofit of steel moment resisting frames subjected to near field ground motions. Earthq. Spectra **17**(3), 427–456 (2001)
Fitzgerald, T.F., Anagnos, T., Goodson, M., Zsutty, T.: Slotted bolted connections in aseismic design for concentrically braced connections. Earthq. Spectra **5**(2), 383–391 (1989)
Freeman, S.A.: Development and use of capacity spectrum method. In: Proceedings of 6th National Conference on Earthquake Engineering, Seattle (1998)
Fu, Y., Kasai, K.: Comparative study of frames using viscoelastic and viscous dampers. J. Struct. Eng. **124**(5), 513–552 (1998)
Grigorian, C.E., Yang, T.S., Popov, E.P.: Slotted bolted connection energy dissipaters. Earthq. Spectra **9**(3), 491–504 (1993)
Hertzberg, R.W.: Deformation and Fracture Mechanics of Engineering Materials. Wiley, New York (1983)
Hwang, J.S.: Seismic design of structures with viscous dampers, 高科技廠房震害防治研討會, 教育部科技顧問室, September. 30, 國立交通大學。, pp. 40–53 (2005)
Hwang, J.S., Huang, Y.N.: Seismic design of structures with viscous dampers, International training program for seismic design of structures, National Center for Research on Earthquake Engineering, 217–234 (2003)
Hwang, J.-S., Huang, Y.-N., Yi, S.-L., Ho, S.-Y.: Design formulations for supplemental viscous dampers to building structures. J. Struct. Eng. **134**(1), 22–31 (2008)
International Code Council (ICC): International building code, 2006th edn. Delmar Cengage Learning, Clifton Park (2006)
Kelly, J.M., Skinner, R.I., Heine, A.J.: Mechanisms of energy absorption in special devices for use in earthquake resistant structures, Bull. N.Z. Soc. Earthq. Eng. **5**(3), 63–88 (1972)
Kim, J., Choi, H.: Displacement-based design of supplemental dampers for seismic retrofit of a framed structure. J. Struct. Eng. ASCE **132**(6), 873–883 (2006)
Kim, J.K., Choi, H.H., Min, K.W.: Performance-based design of added viscous dampers using capacity spectrum method. J. Earthquake Eng. **7**(1), 1–24 (2003)
Kim, J.K., Seo, Y.I.: Seismic design of low-rise steel frames with buckling-restrained braces. Eng. Struct. **26**(5), 543–551 (2004)
Kircher., C.A.: Seismic regulations for passive structural control systems-overview of United States codes. In: Proceedings of 2nd World Conference on Structural Controlo, Wiley, Chichester, U.K (1999)
Lee, D., Taylor, D.P.: Viscous damper development and future trends. Struct. Des. Tall. Build. **10** (5), 311–320 (2001)
Lin, W.H., Chopra, A.K.: Earthquake response of elastic single-degree-of-freedom systems with non-linear viscoelastic dampers. J. Eng. Mech. **129**(6), 597–606 (2003)
Lin, Y.Y., Tsai, M.H., Hwang, J.S., Chang, K.C.: Direct displacement-based design for building with passive energy dissipation systems. Eng. Struct. **25**(1), 25–37 (2003)
Lobo, R.F., Bracci, J.M., Shen, K.L., Reinhorn, A.M., Soong, T.T.: Inelastic response of R/C structures with viscoelastic braces. Earthq. spectra **9**(3), 419–446 (1993)
Makris, N.: Viscous heating of fluid dampers. I: small amplitude motions. J. Eng. Mech. **124**(11), 1210–1216 (1998)
Makris, N., Roussos, Y., Whittaker, A.S., Kelly, J.M.: Viscous heating of fluid dampers. II: large amplitude motions. J. Eng. Mech **124**, 1217–1223 (1998)
Miller, A.K.: Unified Constitutive Equations for Creep and Plasticity. Elsevier Applied Science, London (1987)
Miyamoto, H.K., Determan, L., Gilani, A., Hanson, R.: Seismic rehabilitation of historic concrete structure with fluid viscoelastic dampers. In: Proceedings of 72nd Annual Structural Engineers Association of California Convention, SEAOC, Sacramento, California (2003)
National Fire Protection Association, (NFPA) (2006) NFPA 5000: Building construction and safety code, Quincy, Mass

Nims, D.K., Richter, P.J., Bachman, R.E.: The use of the energy dissipating restraint for seismic hazard mitigation. Earthq. spectra **9**(3), 467–489 (1993)

Ozdemir, H.: Non-linear transient dynamic analysis of yelding structures. Ph. D. Dissertation, University of California, Berkeley, CA (1976)

Pall, A.S., Marsh, C.: Response of friction damped braced frames. J. Struct. Div. ASCE **108** (ST6), 1313–1323 (1982)

Pall, A.S., Marsh, C., Fazio, P.: Friction joints for seismic control of large panel structures. J. Prestressed Concr. Inst. **25**(6), 38–61 (1980)

Pavlou, E., Constantinou, M.C.: Response of elastic andinelastic structures with damping systems to near-field and soft-soil ground motions. Eng. Struct. **26**(9), 1217–1230 (2004)

Ramirez, O.M., Constantinou, M.C., Kircher, C.A., Whittaker, A., Johnson, M., Gomez, J.D., Chrysostomou, C.Z. Development and evaluation of simplified procedures of analysis and design for structures with passive energy dissipation systems. Technical Report No. MCEER-00–0010, Revision, 1, Multidisciplinary Center for Earthquake Engineering Research, University of Buffalo, State University of New York, Buffalo, New York (2001)

Ross, D., Ungar, E.E., Kerwin, E.W.: Damping of plate flexural vibrations by means of viscoelastic laminar structural damping. ASME, New York (1959)

Seleemah, A., Constantinou, M.C.: Investigation of seismic response of buildings with linear and non-linear fluid viscous dampers. Report No. NCEER 97-0004, National Center for Earthquake Engineering Research, State University of New York at Buffalo, Buffalo, New York (1997)

Skinner, R.I., Kelly, J.M., Heine, A.J.: Histeresis dampers for earthquake-resistant structures. Earthq. Eng. Struct. Dynam. **3**, 287–296 (1975)

Soong, T.T., Dargush, G.F.: Passive Energy Dissipation Systems in Structural Engineering. Wiley, Chichester (1997)

Soong, T.T., Spencer, B.F. Jr: Supplemental energy dissipation: state of the art and state of the practice, Eng. Struct. **24**, 243–259 (2002)

Structural Engineers Association of California (SEAOC) (1999) Recommended lateral force requirements and commentary, Appendix I, Sacramento, California

Sullivan, T.J., Calvi, G.M., Priestley, M.J.N., Kowalski, M.J.: The limitations and performances of different displacement based design methods. J. Earthq. Eng. **7**(1), 201–241 (2003)

Symans, M.D., Constantinou, M.C.: Passive fluid viscous damping systems for seismic energy dissipation. ISET J. Earthq. Technol. **35**(4), 185–206 (1998)

Symans, M.D., Charney, F.A., Whittaker, A.S., Constantinou, M.C., Kicher, C.A., Johnson, M. W., McNamara, R.J.: Energy dissipation system for seismic applications: current practise and recent developments. J. Struct. Eng. **134**(1), 3–21 (2008)

Tehrani, P., Maalek, S., The use of passive dampers and conventional strengthening methods for the rehabilitation of an existing steel structure. In: 4th International Conference on Earthquake Engineering Taipei, Taiwan, n.133 (2006)

Tsopelas, P., Okamoto, S., Constantinou, M.C., Ozaki, D., Fujii, S.: NCEER-TAISEI Corporation research program on sliding seismic isolation systems for bridges—Experimental and analytical study of systems consisting of sliding bearings, rubber restoring force devices and fluid dampers. Report No. NCEER 94–0002, National Center for Earthquake Engineering Research, State University of New York at Buffalo, Buffalo, New York (1994)

Wada, A., Saeki, E., Takeuchi, T., Watanabe, A.: Development of Unbonded Brace in Nippon Steel's Unbonded Braces (promotional document), pp. 1–16. Nippon Steel Corporation Building Construction and Urban Development Division, Tokyo (1998)

Wen, Y.K.: Method for random vibration of hysteretic systems. J. Engrg. Mech. Div. **102**(2), 249–263 (1976)

Whittaker, A.S., Aiken, I.D., Bergman, D., Clark, P.W., Cohen, J., Kelly, J.M., Scholl, R.E.: Code requirements for design and implementation of passive energy dissipation systems. In: Proceedings of ATC-17–1 Seminar on Seismic Isolation, Passive Energy Dissipation and Active Control, vol. 2, pp. 497–508. ATC, Redwood City, California (1993)

Whittaker, A.S., Bertero, V., Alonso, J., and Thompson C.: Earthquake simulator testing of steel plate added damping and stiffness elements, Report No. UCB/EERC-89/02, Earthquake Engineering Research Center, University of California, Berkeley (1989)

Whittaker, A.S., Bertero, V.V., Thompson, C.L., Alonso, L.J.: Seismic testing of steel-plate energy dissipating devices. Earthq. Spectra **7**(4), 563–604 (1991)

Xia, C., Hanson, R.D.: Influence of ADAS element parameters on building seismic response, J. Struct. Eng. ASCE **118**(7), 1903–1918 (1992)

Chapter 3
Dynamic Response of Systems Equipped with Viscous and Viscoelastic Dampers

Abstract In this chapter, the principal elements of the dynamic response of linear undamped and damped single and multi degrees-of-freedom systems are described in time domain both in terms of the relative motions of the mass and in state space form. In a first stage, the analysis of the dynamic response of linear simple-degree-of-freedom systems in the time domain is described by defining the main dynamic characteristics of the undamped system and, then, evaluating the damped natural frequency of the system and response factor. The modal strain energy method is described to evaluate the damping ratio of the SDOF system equipped with viscous and viscoelastic devices. As for linear multi-degrees-of-freedom systems the decoupling procedure for modal analysis and the proportional damping model are described. Even for these systems the effect of damping, assumed to be proportional, on the natural frequencies is evaluated and the modal strain energy method with regard to viscous and viscoelastic devices is illustrated. Then, the concept of the state of a system, the definition of the state space and its properties are discussed. Finally, with reference to both single-degree-of-freedom that multi-degrees-of-freedom systems the representation of their dynamic response in state space is illustrated.

3.1 Elements of Dynamic Response of Single-Degree-of-Freedom Systems

The development of new technologies has changed the criteria of seismic design. Nowadays, designing a structure as a static element is absolutely inadequate because the analysis of its dynamic behavior, in fact, has a role of primary importance to obtain specific seismic performance, and, particularly, in the case in

P. Castaldo, *Integrated Seismic Design of Structure and Control Systems*,
Springer Tracts in Mechanical Engineering, DOI: 10.1007/978-3-319-02615-2_3,

which innovative protection strategies are adopted, such as, for example, the introduction of energy dissipation devices within the structural system. For this reason, the principal elements of the dynamic behavior of a single-degree-of-freedom (SDOF) system with specific reference to the effects of damping on the dynamic response are illustrated.

This simple model, even if on the one hand is useful to demonstrate the energy dissipation principles, on the other is not able to adequately model the real structures, for which more complex models need to be introduced, whose dynamic response will be illustrated (Housner 1959; Newmark and Rosenblueth 1971; Soong and Dargush 1999).

Consider the lateral motion of the basic single-degree-of-freedom (SDOF) model, consisting of a mass *m*, supported by springs with total linear elastic stiffness *k*, and a damper with linear viscosity *c*. This SDOF system is then subjected to an excitation seismic input that determines the ground displacements described analytically by law $x_g(t)$. The excited model responds with a lateral displacement x(t) relative to the ground x(t), in such a way that the overall absolute displacement is equal to:

$$x_{ass}(t) = x(t) + x_g(t) \tag{3.1}$$

Since, moreover, both the stiffness and damping properties are linear, the dynamic equation of the system is written as follows:

$$m\ddot{x}(t) + c\dot{x}(t) + kx(t) = -m\ddot{x}_g(t) \tag{3.2}$$

where the mass is the mass of the system including the linear viscous device.

Dividing the Eq. (3.2) by the mass, it applies:

$$\ddot{x}(t) + 2\xi\omega\dot{x}(t) + \omega^2 x(t) = -\ddot{x}_g(t) \tag{3.3}$$

in which the natural (undamped) frequency of the system is defined as:

$$\omega = \sqrt{\frac{k}{m}} \tag{3.4}$$

and ξ is the damping ratio:

$$\xi = \frac{c}{2m\omega} \tag{3.5}$$

In the case of non-zero damping ratio, the free vibration response occurs through the following equation:

$$x(t) = x_0 e^{-\xi\omega t}\left[\cos(\omega_D t) + \frac{\xi\omega}{\omega_D}\sin(\omega_D t)\right] \tag{3.6}$$

where ω_D represents the natural frequency of the damped system defined as follows:

$$\omega_D = \omega\sqrt{1 - \xi^2} \tag{3.7}$$

Note that Eq. (3.7) coincides with the (3.4) in the case $\xi = 0$, the damping ratio is zero, and is valid only in the case of under-damping (i.e., when it is $\xi < 1$). Besides, all the real structures related to the field of seismic engineering are (under)damped.

The dynamic response indicated in Eq. (3.6), is shown in Fig. 3.1, by highlighting the substantial effect due to the increase of damping ratio on the response of the SDOF system. Note, as in the case of $\xi = 0.01$, the system should perform about 37 vibrations before reducing the amplitude to one tenth of its initial value, while for $\xi = 0.20$, the number of vibrations is reduced to 2.

In any case, however, regardless of the value of the damping ratio, the system asymptotically returns to its undeformed configuration, since all the elastic potential energy present at time t = 0 has been dissipated through a purely viscous mechanism.

It should be noted that, in structures equipped with passive supplemental energy dissipation devices, such as viscous dampers, it is good to subdivide the damping ratio in two aliquots:

$$\xi = \xi_s + \xi_d \tag{3.8}$$

where is ξ_s the inherent damping of the main structural system and ξ_d is attributed to the dampers.

The response of the same system to forced vibrations (harmonic excitation) is explained below. Suppose, in fact, that the system is subject to an harmonic excitation, expressed in a complex exponential form, characterized by amplitude F_0 and frequency ω_n:

$$p(t) = F_0 e^{i\omega_n t} \tag{3.9}$$

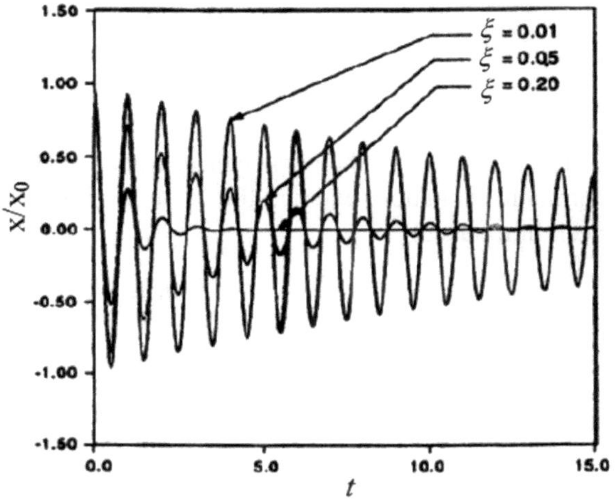

Fig. 3.1 Free vibration response of a (under) damped SDOF system

where i represents the imaginary unit. The dynamic equation of the system appears to be represented, therefore, by the following equation:

$$m\ddot{x}(t) + c\dot{x}(t) + kx(t) = F_0 e^{i\omega_n t} \quad (3.10)$$

The solution of this differential equation turns out to be the sum between the solution of the homogeneous equation associated, representative of free vibrations of the system, and the particular solution.

The solution of the homogeneous equation can be written in a complex exponential form, generalizing Eq. (3.6) as:

$$x_{om}(t) = x_{AD} e^{-\xi\omega t} e^{i(\omega_D t + \phi_D)} \quad (3.11)$$

where x_{AD} and ϕ_D are constants dependent on the initial conditions of motion.

With regard to the particular solution, given the mathematical form of the force applied, it can be expressed such as:

$$x(t) = x_0 e^{i\omega_n t} \quad (3.12)$$

where x_0 is the amplitude of the motion. Substituting the Eq. (3.12) into the equation of motion (3.2) the following expression is obtained:

$$x_0 = \frac{F_0}{\left(-m\omega_n^2 + ic\omega_n + k\right)} \quad (3.13)$$

The complex number, which is the denominator of the Eq. (3.13) can be represented in exponential form through the relationship:

$$k - m\omega_n^2 + ic\omega_n = re^{i\phi_H} \quad (3.14)$$

where, dividing Eq. (3.14) by the mass, and employing the definitions given in Eq. (3.5), the terms r and ϕ_H can be expressed as:

$$\begin{aligned} r &= k\sqrt{\left(1-\beta^2\right)^2 + (2\xi\beta)^2} \\ \phi_H &= arctg\left(\frac{2\xi\beta}{1-\beta^2}\right) \end{aligned} \quad (3.15)$$

In the Eq. (3.15), the parameter β, which is the ratio of the frequency ω_n of the harmonic excitation (external force) to the natural frequency ω of the system, has been introduced:

$$\beta = \frac{\omega_n}{\omega} \quad (3.16)$$

Thus, the Eq. (3.12) can be expressed by the following expression:

$$x_p = x_{st} \frac{e^{-i\phi_H} e^{i\omega_n t}}{\sqrt{\left(1-\beta^2\right)^2 + (2\xi\beta)^2}} \quad (3.17)$$

in which it is possible to introduce the response factor, expressed as:

$$H(\omega_n) = \frac{1}{\sqrt{\left(1-\beta^2\right)^2+(2\xi\beta)^2}} \tag{3.18}$$

The response factor is the ratio of the dynamic to the static displacement response amplitudes, already introduced in Sect. 1.2. Therefore, Eq. (3.17) can be written in compact form as:

$$x_p = x_{st}e^{-i\phi_H}e^{i\omega_n t}H(\omega_n) \tag{3.19}$$

Thus, the overall solution of the problem of motion can be expressed by the formula:

$$x(t) = x_{AD}e^{-\xi\omega t}e^{i(\omega_D t+\phi_D)} + x_{st}e^{-i\phi_H}e^{i\omega_n t}H(\omega_n) \tag{3.20}$$

as usual the arbitrary constants x_{AD} and ϕ_D are determined by the initial conditions of motion.

Observing the Eq. (3.20), it is possible to understand how the first addend is characterized by an "exhaustion" trend (Fig. 3.1), for which it is significant only in the early stages of motion (transitional phase), while after a certain time interval, the motion is virtually identified only by the second addend, which physically represents the stationary phase of the motion.

It is understood, therefore, as in most applications, only the stationary phase of the dynamic response (Eq. 3.20) results to be of practical interest. To characterize the dynamic term, two quantities are essential:

- the response factor $H(\omega_n)$: the dependence of this function on the parameter β for several values of the damping ratio ξ is represented in Fig. 3.2. For small values of the frequency ω_n compared to the frequency of the system ($\omega_n/\omega \approx 0$), the response factor is close to 1, in fact, the response appears to be virtually static since the motion is not quick to point to determine significant inertial effects. In this case, the maximum displacement is controlled by the stiffness of the system with little effect of mass or damping. For low values of damping ratio, with increasing the parameter β the response factor grows rapidly, until it reaches the peak value for β close to unity. For β close to 1, the response factor is controlled by the damping ratio ξ with a negligible influence of mass or stiffness. When $\beta = 1$, the response factor is inversely proportional to the damping ratio if the forcing frequency is the same as the natural frequency of the structure. In this case, the term of resonance is introduced. A resonant frequency is defined as the frequency for which the response is maximum. The response factor $H(\omega_n)$ is essentially independent of damping and approaches zero as the forcing frequency becomes much higher than the natural frequency of the structure ($\omega_n \to \infty$) (i.e., for values of β greater than 1). It can be shown that at high forcing frequencies, the maximum displacement depends primarily on the mass.

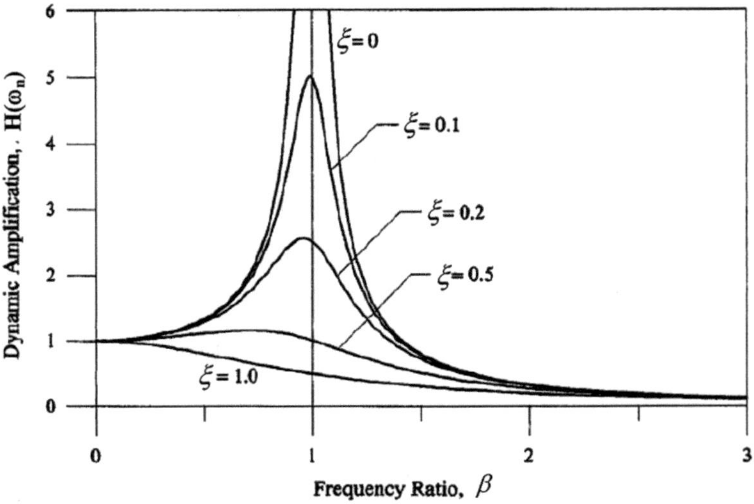

Fig. 3.2 The dependence of the response factor on the fundamental dynamic parameters

- the phase angle ϕ_H: the phase angle is a measure of the delay between the steady-state response of the system and the external force, and is due to the damping ratio. The dependence of the phase angle on the parameter β is represented in Fig. 3.3, for several values of the damping ratio of the system. For undamped systems, the phase angle is equal to 0 in the case $\beta < 1$, and to π for $\beta > 1$, which is equivalent to say that the system response is in phase with the external forcing in the first case, while it is in the opposition phase in the second

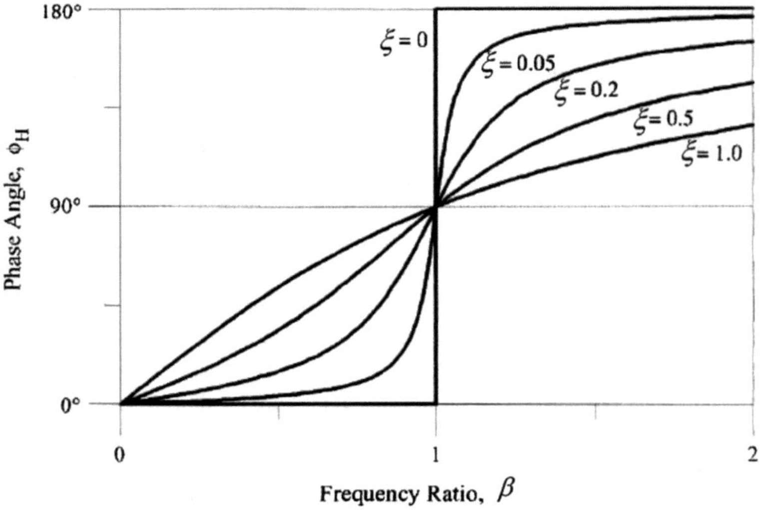

Fig. 3.3 The dependence of the phase angle on the fundamental dynamic parameters

case. For lowly damped structures, the phase angle increases in the range $\left[0, \frac{\pi}{2}\right]$ for $\beta < 1$, up to reach a value equal to $\pi/2$ at resonance. After this point, the phase angle continues to increase until it reaches the value π asymptotically. Note that for a fixed value of the excitation frequency, higher than resonance frequency, the value of the phase angle is inversely proportional to the damping ratio of the system.

3.1.1 Damping Ratio of Single-Degree-of-Freedom Systems Equipped with Linear Viscous Fluid and Viscoelastic Dampers

The evaluation of the dimensionless damping ratio produced by the viscous and/or viscoelastic devices means to calculate the energy dissipated in the hysteresis cycles by the devices.

Considering a single-degree-of-freedom system equipped with a linear viscous damper under an imposed sinusoidal displacement time history (Hwang and Huang 2003; Hwang et al. 2008; Hwang 2005):

$$u = u_0 \sin \omega_n t \tag{3.21}$$

where u is the displacement of the system and the damper; u_0 is the amplitude of the displacement; and the ω_n is the excitation frequency. The measured resistance force is:

$$F = F_0 \sin(\omega_n t + \delta) \tag{3.22}$$

where F is the force response of the system; F_0 is amplitude of the force; and the δ is the phase angle.

The energy dissipated by the damper, W_D, is

$$W_D = \oint F_d du \tag{3.23}$$

where F_d is the damper force which equals to $c\dot{u}$; c is the damping coefficient of the damper; $\dot{u}$ is the velocity of the system and the damper. Therefore, the following is obtained:

$$W_D = \oint c\dot{u}du = \int_0^{2\pi/\omega_n} c\dot{u}^2 dt = cu_0^2\omega_n^2 \int_0^{2\pi} \cos^2 \omega_n t d(\omega_n t) = \pi c u_0^2 \omega_n \tag{3.24}$$

Recognizing that the damping ratio contributed by the damper can be expressed as $\xi_d = c/c_{cr}$, the following is obtained:

$$\begin{aligned} W_D &= \pi c u_0^2 \omega_n = \pi c_{cr} \xi_d u_0^2 \omega_n = 2\pi \xi_d \sqrt{km} \omega_n u_0^2 \\ &= 2\pi \xi_d u_0^2 k \frac{\omega_n}{\omega} = 2\pi \xi_d W_s \frac{\omega_n}{\omega} \end{aligned} \tag{3.25}$$

where c_{cr}, k, m, ω and W_s are respectively the critical damping coefficient, stiffness, mass, natural frequency and elastic strain energy of the system. The damping ratio attributed to the damper can then be expressed as

$$\xi_d = \frac{W_D}{2\pi W_s} \frac{\omega_n}{\omega} \tag{3.26}$$

The dissipated energy W_D and the elastic strain energy W_s are illustrated in Fig. 3.4.

Under earthquake excitations, ω_n is essentially equal to ω, in other terms, the response of the system is characterized by a dynamic response having energy concentrated on its natural frequency (this concept will be also resumed later in Chaps. 4–6), thus giving:

$$\xi_d = \frac{W_D}{2\pi W_s} \tag{3.27}$$

In the case of a viscoelastic device, the viscosity coefficient of the viscoelastic damper, as defined by Eq. (4.12), must be considered to evaluate the dissipated energy and in the calculation of the elastic strain energy must be referred both to the stiffness of the structural system as well as of the device.

It is important to point out that, because of the flexibility of the serially connected braces to the devices, the viscosity coefficient of the viscous device and the stiffness of the brace, as explained in Chap. 4, are dependent on the frequency and, therefore, the corresponding dynamic coefficients must be used in the evaluation of the two above-described energies. Similarly, the viscosity coefficient and stiffness of the viscoelastic device are also dependent on the frequency and, therefore, as described in Chap. 4, the corresponding dynamic terms must be employed. This dependence is even more pronounced when considering the viscoelastic damper-brace component (Chap. 4).

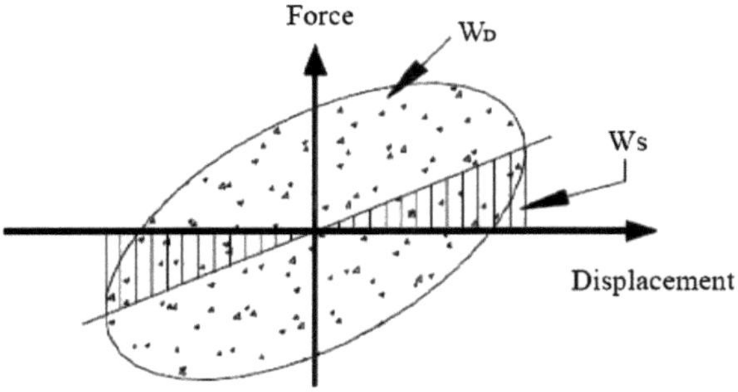

Fig. 3.4 Definition of energy dissipated W_D in a cycle of harmonic motion and maximum strain energy W_S of a SDOF system with viscous damping devices

3.1.2 Damping Ratio of Single-Degree-of-Freedom Systems Equipped with Non Linear Viscous Fluid Dampers

Considering a SDOF system with a non-linear viscous damper (Hwang and Huang 2003; Hwang et al. 2008; Hwang 2005) under sinusoidal motions, the velocity of the system is given by:

$$\dot{u} = \omega_n u_0 \sin \omega_n t \tag{3.28}$$

Recognizing $F_d = cu^{\alpha}$, the energy dissipated by the non-linear damper in a cycle of sinusoidal motion can be expressed as:

$$W_D = \oint F_d du = \int_0^{2\pi/\omega_n} F_d \dot{u} dt = \int_0^{2\pi/\omega_n} \left| c\dot{u}^{1+\alpha} \right| dt = c(u_0 \omega_n)^{1+\alpha} \int_0^{2\pi/\omega_n} \left| \sin^{1+\alpha} \omega_n t \right| dt \tag{3.29}$$

Let $\omega t = 2\theta$ and $dt = 2/\omega d\theta$, using the remarkable Euler-Mascheroni integral, the dissipated energy is written as:

$$\begin{aligned} W_D &= c(u_0\omega_n)^{1+\alpha} \frac{2}{\omega_n} \int_0^{\pi} \left| \sin^{1+\alpha} 2\theta \right| d\theta \\ &= 2^{2+\alpha} c{\omega_n}^{\alpha} u_0^{1+\alpha} \int_0^{\pi/2} 2 \sin^{1+\alpha} \theta \cos^{1+\alpha} \theta d\theta \\ &= 2^{2+\alpha} c{\omega_n}^{\alpha} u_0^{1+\alpha} \frac{\Gamma^2(1+\alpha/2)}{\Gamma(2+\alpha)} \end{aligned} \tag{3.30}$$

where Γ is the Gamma function.

Following a procedure similar to that of the SDOF structure with linear viscous dampers, equivalent damping ratio of the SDOF system contributed by non-linear dampers can be obtained:

$$\xi_d = \frac{\lambda c \omega_n^{\alpha-2} u_0^{\alpha-1}}{2\pi m} \tag{3.31}$$

in which

$$\lambda = 2^{2+\alpha} \frac{\Gamma^2(1+\alpha/2)}{\Gamma(2+\alpha)} \tag{3.32}$$

For the convenience of practical applications, the values of λ are tabulated in FEMA 273 (ATC 1997). It is worth noting that the damping ratio, determined as described above, is dependent on the displacement amplitude u_0.

It is important to point out that, even in this case, the viscosity coefficients of the devices, which define the dimensionless viscous damping ratio, are dependent on the frequency as explained in Chap. 4.

3.2 Elements of Dynamic Response of Multi-Degrees-of-Freedom Systems

The structural systems, unlike the SDOF system analysed in the previous section, may have a large number N of kinematic parameters (degrees of freedom) in terms of generalized relative displacements, which can be considered as the components of a vector **x**(t) of dimension N. In so doing, in perfect analogy with the Eq. (3.2), the N equations of motion of the linear multi-degrees-of-freedom system (MDOF), equipped with viscous linear devices and subject to earthquake ground excitation, can be written in matrix form as (Smith et al. 1992):

$$(\mathbf{M}+\bar{\mathbf{M}})\ddot{\mathbf{x}}(t)+(\mathbf{C}+\bar{\mathbf{C}})\dot{\mathbf{x}}(t)+(\mathbf{K}+\bar{\mathbf{K}})\mathbf{x}(t)=-(\mathbf{M}+\bar{\mathbf{M}})\ddot{\mathbf{x}}_g(t) \qquad (3.33)$$

where **M**, **C** and **K** represent respectively the mass, damping and stiffness matrices of the main system and, $\bar{\mathbf{M}}$, $\bar{\mathbf{C}}$ and $\bar{\mathbf{K}}$ represent respectively the mass, damping and stiffness matrices of the bracing-linear viscous/viscoelastic damper system.

In addition, the vector $\ddot{\mathbf{x}}_g(t)$ represents the contribution of the seismic excitation related to every degree of freedom.

By defining the following matrices:

$$\begin{aligned}\widehat{\mathbf{M}}&=\mathbf{M}+\bar{\mathbf{M}}\\ \widehat{\mathbf{C}}&=\mathbf{C}+\bar{\mathbf{C}}\\ \widehat{\mathbf{K}}&=\mathbf{K}+\bar{\mathbf{K}}\end{aligned} \qquad (3.34)$$

the dynamic matrix equation of the system can be rewritten in the form:

$$\widehat{\mathbf{M}}\,\ddot{\mathbf{x}}(t)+\widehat{\mathbf{C}}\,\dot{\mathbf{x}}(t)+\widehat{\mathbf{K}}\mathbf{x}(t)=-\widehat{\mathbf{M}}\ddot{\mathbf{x}}_g(t) \qquad (3.35)$$

The dynamic matrix Eq. (3.35) is written in the classical form of structural dynamic analysis, which, in the simplest case, presents constant-coefficients and time-invariant matrices. Therefore, it represents, in general, a constant-coefficients system of N coupled second-order differential equations. This mathematical property considerably complicates the resolution of the equation unless the equations are not uncoupled by transformation, or in other words if $\hat{\mathbf{M}}$, $\hat{\mathbf{C}}$ and $\hat{\mathbf{K}}$ can be diagonalized by the same basis of vectors. This condition does not occur for arbitrary values of the matrices involved, but can be obtained by making particular assumptions relatively to the damping matrix, as it will be discussed later.

Consider the free vibrations response of undamped multi-degrees-of-freedom (MDOF) systems, the computation of the vibration properties requires solution of the matrix Eq. (3.36), which in mathematical terminology defines an eigen-problem:

$$(\hat{\mathbf{K}} - \omega^2 \hat{\mathbf{M}})\boldsymbol{\varphi} = \mathbf{0} \tag{3.36}$$

where ω represents a natural vibration frequency of the structure including the braces and without any damping, and $\boldsymbol{\varphi}$ the vector of modal shape associated with it. The resolution of the problem leads to the N natural frequencies and modes of vibration, denoted respectively by ω_i and φ_i for $i = 1, \ldots, N$, starting from the one with a lower frequency (ω_1), defined fundamental frequency. The modes are orthogonal with respect to the mass and stiffness matrices:

$$\begin{aligned} \phi_i^T \hat{\mathbf{M}} \phi_j &= \begin{cases} 1 \; per i = j \\ 0 \; per i \neq j \end{cases} \\ \phi_i^T \hat{\mathbf{K}} \phi_j &= \begin{cases} \omega_i^2 \; per i = j \\ 0 \; per i \neq j \end{cases} \end{aligned} \tag{3.37}$$

in addition to representing a basis of the vector space $\Re^n$.

For notational convenience, the natural frequencies will be the diagonal elements of a matrix denoted by $\boldsymbol{\Omega}^2$, while the vectors of the corresponding modes will be the columns of a square matrix denoted by $\boldsymbol{\Phi}$, which will allow to make a modal transformation of the coordinates from normal coordinates (i.e., displacements at the nodes) to modal coordinates (i.e., amplitudes of the natural mode shapes). Each vector of the relative displacements $\mathbf{x}$ can be represented, therefore, as:

$$\mathbf{x} = \boldsymbol{\Phi}\mathbf{y} \tag{3.38}$$

where $\mathbf{y}$ is the vector of modal coordinates, also called main coordinates of motion. Substituting the expression (3.38) in the Eq. (3.35), through simple mathematical operations, the following equation of motion expressed in terms of principal coordinates is obtained:

$$\ddot{\mathbf{y}} + \boldsymbol{\Phi}^{\mathrm{T}} \hat{\mathbf{C}} \boldsymbol{\Phi} \dot{\mathbf{y}} + \boldsymbol{\Omega}^2 \mathbf{y} = -\ddot{\mathbf{y}}_g \tag{3.39}$$

where $\ddot{\mathbf{y}}_g$ results being:

$$\ddot{\mathbf{y}}_g = \boldsymbol{\Phi}^{\mathrm{T}} \hat{\mathbf{M}} \ddot{\mathbf{x}}_g \tag{3.40}$$

In general, the Eq. (3.40) still represents a system of coupled differential equations, the system, in fact, is uncoupled only if the matrix $\boldsymbol{\Phi}^{\mathbf{T}}\mathbf{C}\boldsymbol{\Phi}$ is diagonal. This happens, for example, in the case of proportional modal damping, or Rayleigh damping, in which it is possible to identify two constants α_0 and α_1 so that the damping matrix can be written as:

$$\hat{\mathbf{C}} = \alpha_0 \hat{\mathbf{M}} + \alpha_1 \hat{\mathbf{K}} \tag{3.41}$$

by exploiting the orthogonality property, the following is given:

$$\mathbf{\Phi}^{\mathrm{T}} \hat{\mathbf{C}} \mathbf{\Phi} = \alpha_0 \mathbf{I} + \alpha_1 \mathbf{\Omega}^2 \tag{3.42}$$

which turns out to be a diagonal matrix.

In case the damping matrix was not proportional, the approach shown can be generalized through the use of the following series of Caughey (Caughey and Ma 1993):

$$\hat{\mathbf{C}} = \hat{\mathbf{M}} \sum_{j=0}^{N-1} \alpha_j \left[\hat{\mathbf{M}}^{-1} \hat{\mathbf{K}} \right]_j \tag{3.43}$$

that allows to carry out also the diagonalization in the general case.

It can also define the damping ratio corresponding to the i-th mode ξ_i, defining a diagonal matrix $\boldsymbol{\xi}$ whose elements are precisely ξ_i, and such as to satisfy the relation:

$$\mathbf{\Phi}^{\mathrm{T}} \hat{\mathbf{C}} \mathbf{\Phi} = 2\xi\Omega \tag{3.44}$$

Substituting the Eq. (3.44) in the Eq. (3.39), the following is obtained:

$$\ddot{\mathbf{y}} + 2\boldsymbol{\xi}\mathbf{\Omega}\dot{\mathbf{y}} + \mathbf{\Omega}^2 \mathbf{y} = -\ddot{\mathbf{y}}_g \tag{3.45}$$

that represents a set of N differential uncoupled equations. In particular, the i-th equation of the system takes the following form:

$$\ddot{y}_i + 2\xi_i \omega_i \dot{y}_i + \omega_i^2 y_i = -\ddot{y}_{g,i} \tag{3.46}$$

that is exactly the dynamic equation of a SDOF system.

Therefore, all the considerations related to the single-degree-of-freedom system can be applied to a single mode of a MDOF system. In fact, the response of a MDOF system can be seen as the combination of the single responses of the "N" SDOF systems corresponding to the "N" mode shapes. To return to the vector of displacement of the system from the modal coordinates, known functions $y_i(t)$, the relation (3.38) can be simply applied.

In the presented process, the more computational effort is the calculation of mode shapes and frequencies of vibration of the system. In this context, however, the consideration relating to the fact that generally only few modes significantly contribute to the motion of the system, simplifies the analysis since only some of the possible modes must be identified.

The modal approach allows to solve the problem in a particularly simple way. This is possible due to the simplifying assumptions made, with particular reference to the following hypotheses:

- The matrices involved in the analysis have constant coefficients; this does not allow to capture the variations in terms of stiffness, mass and damping related to the effect of dynamic load;
- The damping matrix must be diagonalizable, which imposes a strong constraint on the distribution and values of the damping coefficients of viscous devices.

These assumptions appear to be sometimes too restrictive, that is the reason why the principles of representation in the "state space" of dynamic response of a system will be shown later. This representation, in fact, allows to describe the free response of the system for its entire state through a modal approach, which also allows for a representation of the elements of the velocity vector corresponding to every degree of freedom.

3.2.1 Damping Ratio of Multi-Degrees-of-Freedom Systems Equipped with Linear Viscous Fluid and Viscoelastic Dampers

As regards MDOF systems equipped with linear viscous or viscoelastic passive dissipation devices, the motion equations expressed in the modal coordinates remain, as previously stated, always uncoupled in the case of proportional or Rayleigh damping. The effects on the mode damping and stiffness due to the addition of the viscous or viscoelastic devices and the braces can be obtained by following the modal strain energy method that is presented below with reference to the possibility of the use of viscous and/or viscoelastic devices (Ungar and Kerwin 1962; Johnson and Kienholz 1982; Soong and Lai 1991).

The damping ratio for the i-th mode due to the addition of viscoelastic devices can be evaluated as (Soong and Dargush 1997):

$$\xi_{d,i} = \bar{\xi}_i = \frac{\eta(\omega_i')E_v}{2E_i} \tag{3.47}$$

where ω_i' is the frequency of the i-th mode of the structure without devices and braces, $\eta(\omega_i')$ is the loss factor of the viscoelastic material as a function of mode frequency of the original structure (as explained in Chap. 4) relating to the terms of the damping matrix $\bar{\mathbf{C}}$, E_i is the strain energy relative to the i-th mode shape of the system with the devices and E_v is the energy stored in viscoelastic devices. These energies are evaluated as:

$$E_v = \phi_i^T \bar{\mathbf{K}} \phi_i \tag{3.48}$$

$$E_i = \phi_i^T (\mathbf{K} + \bar{\mathbf{K}}) \phi_i \tag{3.49}$$

where ϕ_i is the vector of the form corresponding to the i-th mode shape associated with ω_i', **K** is the stiffness matrix of the original structure, and $\bar{\mathbf{K}}$, according to what is said in the previous section, is the stiffness matrix attributed to the devices.

The Eq. (3.47) can be written as:

$$\xi_{d,i} = \bar{\xi}_i = \frac{\eta(\omega_i')}{2}\left[\frac{\phi_i^T\bar{\mathbf{K}}\phi_i}{\phi_i^T(\mathbf{K}+\bar{\mathbf{K}})\phi_i}\right] = \frac{\eta(\omega_i')}{2}\left[1-\frac{\phi_i^T\mathbf{K}\phi_i}{\phi_i^T(\mathbf{K}+\bar{\mathbf{K}})\phi_i}\right] \tag{3.50}$$

The frequency of the i-th mode of the structure with viscoelastic devices and braces, being reasonably good approximation $\mathbf{M} \approx \hat{\mathbf{M}} = \mathbf{M} + \bar{\mathbf{M}}$, is:

$$\omega_i = \left[\frac{\phi_i^T(\mathbf{K}+\bar{\mathbf{K}})\phi_i}{\phi_i^T\mathbf{M}\phi_i}\right] \tag{3.51}$$

where **M** is the mass matrix of the structure. The Eq. (3.50) can then be simplified as:

$$\xi_{d,i} = \bar{\xi}_i = \frac{\eta(\omega_i')}{2}\left[1-\frac{\phi_i^T\mathbf{K}\phi_i}{\phi_i^T(\mathbf{K}+\bar{\mathbf{K}})\phi_i}\frac{\phi_i^T\mathbf{M}\phi_i}{\phi_i^T\mathbf{M}\phi_i}\right] = \frac{\eta(\omega_i')}{2}\left(1-\frac{\omega_i'^2}{\omega_i^2}\right) \tag{3.52}$$

With regard to structures equipped with viscoelastic devices, the most important design parameter is the damping ratio.

Therefore, evaluating the mode frequencies and damping ratios is an iterative procedure. In the situation in which the modulus of the material corresponding to the frequency ω_i is substantially different from that relating to $\bar{\omega}_i$, it is necessary that the above steps are repeated for a better estimate of the mode quantities. However, it has been highlighted that the iteration process is normally short (Soong and Constantinou 1994; Soong and Dargush 1997).

A similar approach can be carried out with reference to a MDOF system equipped with viscous devices, as shown in Fig. 3.5.

Considering a MDOF system (Hwang and Huang 2003; Hwang et al. 2008; Hwang 2005) equipped with viscous devices, the total effective damping ratio of the system, ξ, is defined as:

$$\xi = \xi_s + \xi_d \tag{3.53}$$

where ξ_s is the inherent damping ratio of the MDOF system without dampers, and ξ_d is the viscous damping ratio attributed to added dampers. By extending the theory related to a SDOF system, the equation shown below is used by FEMA273 (ATC 1997) to evaluate ξ_d:

$$\xi_d = \frac{\sum W_j}{2\pi W_k} \tag{3.54}$$

where ΣW_j is the sum of the energy dissipated by the j-th damper of the system in one cycle; and W_K is the elastic strain energy of the frame. W_K is equal to $\Sigma F_i\Delta_i$

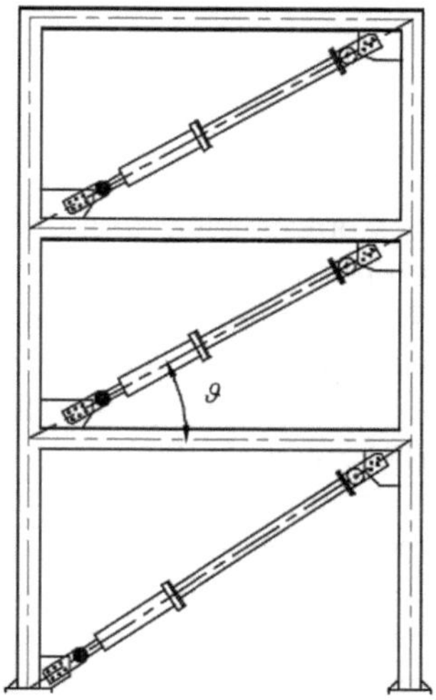

Fig. 3.5 MDOF structural system equipped with viscous dampers

where F_i is the story shear and Δ_i is the story drift of the i-th floor. Now, the energy dissipated by the viscous dampers can be expressed as:

$$\sum_j W_j = \sum_j \pi c_j u_j^2 \omega = \frac{2\pi^2}{T} \sum_j c_j u_j^2 \tag{3.55}$$

where u_j is the relative axial displacement of j-th damper between the two ends. Experimental evidence (Hwang and Huang 2003; Hwang et al. 2008; Hwang 2005) has shown that if the damping ratio of a structure is increased the higher mode responses of the structure will be suppressed. As a consequence, only the first mode of a MDOF system is usually considered in the simplified procedure of practical applications. Using the modal strain energy method, the energy dissipated by the dampers and the elastic strain energy provided by the primary frame can be rewritten as:

$$\sum_j W_j = \frac{2\pi^2}{T_1} \sum_j c_j \phi_{dj}^2 \cos^2 \vartheta_j \tag{3.56}$$

$$W_k = \Phi_1{}^T \mathbf{K} \Phi_1 = \Phi_1{}^T \omega_1{}^2 \mathbf{M} \Phi_1 = \sum_i \omega_1{}^2 m_i \phi_i{}^2 = \frac{4\pi^2}{T_1{}^2} \sum_i m_i \phi_i{}^2 \tag{3.57}$$

where **K**, **M**, Φ_1 are respectively the stiffness matrix, the mass matrix and the first mode of the system; ϕ_{dj} is the relative horizontal displacement of j-th damper corresponding to the first mode shape; ϕ_i is the first mode displacement at i-th floor; m_i is the mass of i-th floor; and θ_j is the inclined angle of j-th damper. Substituting the Eqs. (3.54)–(3.57) into the expression (3.53), the effective damping ratio of a structure with linear viscous dampers is given by:

$$\xi = \xi_s + \frac{\frac{2\pi^2}{T_1}\sum_j c_j\phi_{rj}^2\cos^2\vartheta_j}{2\pi\frac{4\pi^2}{T_1^2}\sum_i m_i\phi_i^2} = \xi_s + \frac{T_1\sum_j c_j\phi_{rj}^2\cos^2\vartheta_j}{4\pi\sum_i m_i\phi_i^2} \tag{3.58}$$

Corresponding to a desired added damping ratio, there is no substantial procedure suggested by design codes for distributing c values over the whole building. When designing the dampers, it may be convenient to distribute the c values equally in each floor. However, many experimental results have shown that the efficiency of dampers on the upper stories is smaller than that in the lower stories (Pekcan et al. 1999).

It must be pointed out that, in a more complete analysis, the contribution due to the viscoelastic dissipation system both in terms of damping as well as stiffness corresponding to each mode depends on the frequency of the relative mode. The next chapter deals with the dynamic response of both the viscoelastic damper-brace component and the viscous damper-brace component by taking into account the stiffness of the brace. It follows that to evaluate the terms of the matrices $\bar{\mathbf{k}}(\omega)$ and $\bar{\mathbf{C}}(\omega)$, the dynamic values related to the stiffness and viscosity of the "viscoelastic or viscous device-brace component" must be considered.

Further, it should also be noted that in the case of high damping, the damped frequency corresponding to every mode must be taken into account:

$$\omega_{D,i} = \omega_i\sqrt{1 - \xi_i^2} \tag{3.59}$$

With reference to this problem, in the case of a dimensionless damping ratio higher than 20 %, by following the modal strain energy method, a modified approach in the complex frequencies space has been proposed (Tsai and Chang 2000, 2001).

3.2.2 Damping Ratio of Multi-Degrees-of-Freedom Systems Equipped with Non Linear Viscous Fluid Dampers

For a MDOF system with non-linear dampers (Hwang and Huang 2003; Hwang et al. 2008; Hwang 2005), the damping ratio can be obtained as follows. Considering the first mode only, the elastic strain energy is:

$$W_k = \omega_1^2\sum_i m_i u_i^2 \tag{3.60}$$

Assuming that all the dampers of the whole building have the same α and substituting the Eqs. (3.24), (3.25) and (3.21) into (3.27), the damping ratio contributed by the dampers is obtained as:

$$\xi_d = \frac{\sum_j \lambda c_j u_{dj}^{1+\alpha} \cos^{1+\alpha} \vartheta_j}{2\pi\omega^{2-\alpha} \sum_i m_i u_i^2} \tag{3.61}$$

where u_{dj} is the relative displacement between the ends of j-th damper in the horizontal direction. Since only the first mode is considered, the displacement response may be expressed as:

$$u_i = A\phi_i \tag{3.62}$$

where ϕ_i is the first modal displacement of the i-th degree-of-freedom and A is the amplitude.

Finally, substituting Eqs. (3.22) and (3.23) to (3.27), the effective damping ratio of the structure with non-linear dampers is obtained:

$$\xi = \xi_s + \frac{\sum_j \lambda c_j \phi_{dj}^{1+\alpha} \cos^{1+\alpha} \vartheta_j}{2\pi A^{1-\alpha} \omega^{2-\alpha} \sum_i m_i \phi_i^2} \tag{3.63}$$

Even in this case, the presence of the braces should not be neglected and, consequently, the dynamic values of viscosity of each non-linear viscous device and the stiffness of each brace should be employed.

3.3 Elements of Dynamic Response of Linear Systems in the State Space

Linear dynamic systems can be studied in the time or frequency domain. The first representation allows to obtain the time response of a system having fixed dynamic characteristics, instead, the second representation allows to evaluate the response of a multitude of systems with different dynamic characteristics.

A very important function and often used in the representations described above is the transfer function, defined as follows:

$$G(s) = \frac{Y(s)}{X(s)} = \frac{b_m s^m + b_{m-1} s^{m-1} + \ldots + b_1 s + b_0}{s^n + a_{n-1} s^{n-1} + a_{n-2} s^{n-2} + \ldots + a_1 s + a_0} \tag{3.64}$$

Zeros of polynomials Y(s) and X(s) are called, respectively, zeros and poles of the transfer function (or the system). A typical graphical representation of a dynamic Single Input Single Output (SISO) system through the transfer function is shown in Fig. 3.6.

The transfer function *G*(*s*) is in bijective correspondence with the following differential equation in which *x*(*t*) is the input function and *y*(*t*) the output one:

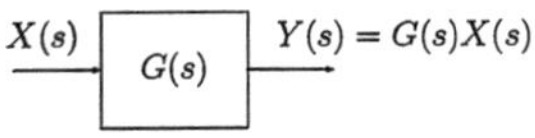

Fig. 3.6 Graphical representation of the dynamic SISO system

$$y^{(n)}(t) + \ldots + a_1\dot{y}(t) + a_0y(t) = b_mx^{(m)}(t) + \ldots + b_1\dot{x}(t) + b_0x(t) \tag{3.65}$$

With $y^{(n)}(t)$ is indicated the n-th derivative of the temporal function $y(t)$. Note that the transfer function $G(s)$ is used only for linear constant-coefficients and time-invariant systems.

The bijective correspondence means that at any time it must be possible to pass by the differential equation to the transfer function and the other way round, assuming that the initial conditions are all identically zero.

One way to write the linear dynamic systems is to use a representation in the state space. In this case, the dynamic system can be described using a number of variables (typically internal variables) which defines the state of the system. The values assumed by the state variables at a generic instant of time contain, in their entirety, all the information on the past response of the system useful in order to evaluate the future response knowing the input variable values at following instants of time.

To determine the state $x(t)$ at time t, it is necessary to know the state $x(t_0)$ at time t_0 and the input function $u(\cdot)$ in the time interval $[t_0, t]$ (Fig. 3.7).

$$x(t) = \psi(t, t_0, x_0, u(\cdot)) \tag{3.66}$$

The concept of the state does not take into account the past history of the system prior to the instant t_0.

A linear time invariant system is described in the state space form as follows:

$$\begin{cases} \dot{\mathbf{x}}(t) = \mathbf{Ax(t)} + \mathbf{Bu(t)} \\ \mathbf{y(t)} = \mathbf{Cx(t)} + \mathbf{Du(t)} \end{cases} \tag{3.67}$$

where with $\mathbf{x}(t) = [x_1x_2...x_n]^T \in \Re^n$ the state vector is indicated, with $\mathbf{u}(t) \in \Re^m$ the input vector, with $\mathbf{y}(t) \in \Re^p$ the output vector and with $\mathbf{A} \in \Re^{n\times n}$, $\mathbf{B} \in \Re^{n\times m}$, $\mathbf{C} \in \Re^{p\times n}$, $\mathbf{D} \in \Re^{p\times m}$ matrices of appropriate dimension.

3.4 Motion Equation of Dynamic Response of Single-Degree-of-Freedom Systems in the State Space Form

In the previous sections, the dynamic response of SDOF and MDOF systems has been illustrated by solving a constant-coefficients system of second order differential equations: "modal analysis". An alternative approach is to describe the dynamic response in the state space form.

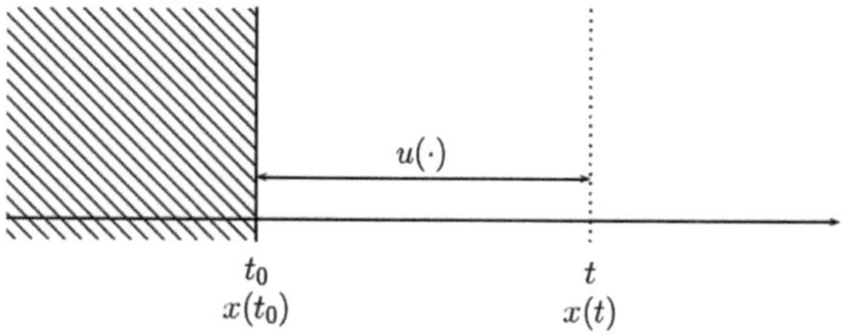

Fig. 3.7 Physical meaning of the state variables

Consider the dynamic Eq. (3.2) related to a SDOF system. The integration of this equation, once having established the initial conditions of motion, leads to the determination of the velocity and displacement for each instant of time. Note that these two variables allow to value through the same Eq. (3.2), the acceleration of the system and, therefore, all the dynamic actions on the system. In other words, the two variables are sufficient for the complete description of the system and define, therefore, "the dynamic state".

Thus, rather than describing the dynamic response by a differential second order equation, it is convenient to transform the Eq. (3.2) into a system of two differential first order equations which contain as unknowns, the state variables displacement, x(t), and velocity, $\dot{x}(t)$. The new form of the problem is as follows:

$$\begin{cases} \dfrac{dx}{dt} = \dot{x} \\ \dfrac{d\dot{x}}{dt} = \left(-\dfrac{c}{m}\right)\dot{x} + \left(-\dfrac{k}{m}\right)x - \ddot{u}_g \end{cases} \tag{3.68}$$

This representation is called the "state space form" and presents a reduced complexity on the research of analytical and numerical solutions of the problem.

To write the Eq. (3.68) by means of a matrix equation, the state vector **z** is defined, with the variables that represent the state of the system as components:

$$\mathbf{z}(t) = \begin{bmatrix} x(t) \\ \dot{x}(t) \end{bmatrix} \tag{3.69}$$

in such a way that the Eq. (3.68) can be represented through the following matrix equation:

$$\dot{\mathbf{z}}(t) = \mathbf{A}\mathbf{z}(\mathbf{t}) + \mathbf{b}\ddot{u}_g(t) \tag{3.70}$$

in which:

$$\mathbf{A} = \begin{bmatrix} 0 & 1 \\ -k/m & -c/m \end{bmatrix} \qquad \mathbf{b} = \begin{bmatrix} 0 \\ 1 \end{bmatrix} \tag{3.71}$$

the matrix **A** is generally defined by the term "status matrix". Matrix **A** and vector **b**, by introducing the magnitudes, presented in Eqs. (3.4) and (3.5), frequency ω and damping factor ξ, can be written:

$$\mathbf{A} = \begin{bmatrix} 0 & 1 \\ -\omega^2 & -2\xi\omega \end{bmatrix} \quad \mathbf{b} = \begin{bmatrix} 0 \\ 1 \end{bmatrix} \tag{3.72}$$

3.5 Motion Equation of Dynamic Response of Multi-Degrees-of-Freedom Systems in the State Space Form

With reference to MDOF systems, it is possible to represent the matrix Eq. (3.33) in the state space form. The state of the MDOF system will be represented by a vector of dimension 2N containing the displacements and velocities in relation to each of the N degrees of freedom:

$$\mathbf{z(t)} = \begin{bmatrix} \mathbf{u(t)} \\ \dot{\mathbf{u}}\mathbf{(t)} \end{bmatrix} \tag{3.73}$$

while the matrix **A** and the vector **b** will take the following form:

$$\mathbf{A} = \begin{bmatrix} \mathbf{0}_{N\times N} & \mathbf{I}_{N\times N} \\ -\mathbf{M}^{-1}\mathbf{K} & -\mathbf{M}^{-1}\mathbf{C} \end{bmatrix} \quad \mathbf{b} = \begin{bmatrix} \mathbf{0}_{\mathrm{N}\times 1} \\ \mathbf{i} \end{bmatrix} \tag{3.74}$$

in which the terms $\mathbf{0}_{\mathrm{N} \times \mathrm{N}}$, $\mathbf{0}_{\mathrm{N} \times 1}$, $\mathbf{I}_{\mathrm{N} \times \mathrm{N}}$ and **i** represent, respectively, the zero matrix of size N × N, the zero column vector of size N, the identity matrix of size N × N and the vector of influence representative of the seismic action on the structure. The matrices **M**, **K** and **C** contain the sum of the terms related to the main system and the dissipative bracing-damper system.

The Eq. (3.70), therefore, for a N degrees-of-freedom structure is mathematically a system of 2N differential first order equations. The solution of this differential matrix equation is illustrated below describing in advance the free vibrations response of the structural system.

3.5.1 Free Vibration Response

A possible solution of the problem of free vibrations of the system can be obtained through the following relation:

$$\mathbf{z} = \mathbf{v}e^{\lambda t} \tag{3.75}$$

where **v** is a generic vector of dimension 2N and λ is a scalar.

Substituting Eq. (3.75) into (3.70), the following equation is obtained:

$$(\mathbf{A} - \lambda\mathbf{I}_{2N\times 2N})\mathbf{v} = \mathbf{0}_{2N\times 2N} \tag{3.76}$$

in which $\mathbf{I}_{2N\times 2N}$ and $\mathbf{0}_{2N\times 2N}$ are respectively the identity matrix and zero matrix of dimension 2N × 2N.

In order that the Eq. (3.76) presents non-trivial solutions, it must be:

$$\det(\mathbf{A} - \lambda\mathbf{I}_{2N\times 2N}) = \det\begin{bmatrix} -\lambda\mathbf{I}_{N\times N} & \mathbf{I}_{N\times N} \\ -\mathbf{M}^{-1}\mathbf{K} & -\mathbf{M}^{-1}\mathbf{C} - \lambda\mathbf{I}_{N\times N} \end{bmatrix} = 0 \tag{3.77}$$

which is an algebraic second degree equation in the variable λ, and is mathematically the eigen-problem to find the eigenvalues of the state matrix. Since, in general, structural systems are under-damped, the Eq. (3.77) presents conjugate imaginary solutions representative of the natural frequencies of vibration of the N modes of the structure.

It will be indicated, therefore, with λ_i the eigenvalue corresponding to the i-th mode of vibration, and with $\tilde{\lambda}_i$ its conjugate.

The modal coordinates corresponding to each of the eigenvalues are determined by substituting the latter in Eq. (3.76). It seems evident that the eigenvectors corresponding to the conjugate imaginary eigenvalues are conjugate imaginary vectors. The generic eigenvector will be indicated with $\mathbf{v}_i$ and with $\tilde{\mathbf{v}}_i$, its conjugate. Note that the eigenvectors are characterized by a number of components equal to 2N, they, in fact, are representative of the mode shapes relative to the state vector, both in terms of displacement and velocity.

The solution of the associated homogeneous equation can thus be expressed by the formula:

$$\mathbf{z}_{om} = \sum_{i=1}^{N}\left(C_i e^{\lambda_i t}\mathbf{v}_i + C_i' e^{\tilde{\lambda}_i t}\tilde{\mathbf{v}}_i\right) \tag{3.78}$$

where C_i and C_i' can be written as follows:

$$C_i = \frac{1}{2}\left(C_R^i + iC_I^i\right) \tag{3.79}$$

$$C_i' = \tilde{C}_i \tag{3.80}$$

and C_R^i, C_I^i for i = 1…N are 2N arbitrary real constants to be defined according to the initial conditions of motion.

Note that the modal analysis within the state space does not require a particular form of the damping matrix **C**, which also allows to describe the non proportional distribution of the dissipative resources.

3.5.2 Response to Harmonic Excitation

In the case of a system subjected to an external force, a possible solution by using the convolution integral of Duhamel is provided.

Consider, therefore, the following scalar differential first order equation:

$$\dot{y}(t) = ay(t) + g(t) \tag{3.81}$$

the total solution of this problem can be expressed in the following form:

$$y(t) = e^{a(t-t_0)}y_0 + \int_{t_0}^{t} e^{a(t-\tau)}g(\tau)d\tau \tag{3.82}$$

where y_0 is the value of variable y at the instant t_0.

With reference to the problem expressed by the Eq. (3.70), it is possible to obtain a solution which has a similar form. Regarding the solution of the associated homogeneous equation, it is possible to demonstrate, in fact, how the Eq. (3.78) can be written in a compact form as follows:

$$\mathbf{z}_{om}(t) = e^{\mathbf{A}t}\mathbf{z}_0 \tag{3.83}$$

in which $e^{\mathbf{A}t}$, called as "transition matrix", is a matrix defined by the following series:

$$e^{\mathbf{A}t} = \mathbf{I} + \mathbf{A}t + \frac{1}{2}\mathbf{A}\mathbf{A}t^2 + \ldots + \frac{1}{n!}\mathbf{A}^n t^n + \ldots \tag{3.84}$$

The fundamental property of this matrix is that it presents a time derivative entirely similar to that of the corresponding scalar function:

$$\frac{de^{\mathbf{A}t}}{dt} = \mathbf{A}e^{\mathbf{A}t} \tag{3.85}$$

By using the Eq. (3.83) and the integral of Duhamel, it is possible to provide a general expression of the solution (3.70):

$$\mathbf{z}(t) = e^{\mathbf{A}(t-t_0)}\mathbf{z}_0 + \int_{t0}^{t} e^{\mathbf{A}(t-\tau)}\mathbf{b}\ddot{u}_g(\tau)d\tau \tag{3.86}$$

The Eq. (3.86) is particularly useful from a numerical point of view because it simply allows the numerical integration of the problem through the discretization of the time variable.

In this chapter, the main elements of the dynamic analysis of damped systems with viscous devices have been provided. The issues covered, far from being a complete discussion of the problem, have the aim of giving the main concepts related to damped systems in order to understand what is described in the next chapters.

The dynamic analysis in the state space allows for an assessment of the response of the structure that is not bound by particularly restrictive hypothesis and describes the free response of the system through a modal approach relative to the entire state of the system, which allows for a representation of the velocity vector corresponding to the single degree of freedom.

References

Applied Technology Council (ATC): NEHRP commentary on the guidelines for the seismic rehabilitation of buildings. Report No. FEMA-273, prepared for the Building Seismic Safety Council (BSSC) by the ATC, Federal Emergency Management Agency (FEMA), Washington, DC (1997)

Caughey, T.K., Ma, F.: Complex modes and solvability of non classical linear systems. J. Appl. Mech. **32**, 583–588 (1993)

Housner, G.W.: Behaviour of structures during earthquake. J Mech. Eng. **85**, 109–129 (1959)

Hwang, J.S.: Seismic design of structures with viscous dampers, 高科技廠房震害防治研討會, 教育部科技顧問室. 國立交通大學。, Sept 30, pp. 40–53 (2005)

Hwang, J.S., Huang, Y.N.: Seismic design of structures with viscous dampers, International training program for seismic design of structures, National Center for Research on Earthquake Engineering, pp. 217–234 (2003)

Hwang, J.-S., Huang, Y.-N., Yi, S.-L., Ho, S.-Y.: Design formulations for supplemental viscous dampers to building structures. J Struct. Eng. **134**(1), 22–31 (2008)

Johnson, C.D., Kienholz, D.A.: Finite element prediction of damping in structures with constrained viscoelastic layers. AIA A J **20**(9), 1284–1290 (1982)

Newmark N.M., Rosenblueth E.: Fundamentals of Earthquake Engineering, Prentice Hall, Englewood Cliffs, New Jersey (1971)

Pekcan, G., Mander, J.B., Chen, S.S.: Design and retrofit methodology for building structures with supplemental energy dissipating systems, Report No MCEER 99-0021. Multidisciplinary Center for Earthquake Engineering Research, Buffalo (1999)

Smith, M.J., Grigoriadis, K.M., Skelton, R.: Optimal mix of passive and active control in structures. J. Guide Control Dyn. **15**(4), 912–919 (1992)

Soong, T.T., Constantinou, M.C.: Passive and Active Structural Vibration Control in Civil Engineering. Springer, NY (1994)

Soong, T.T., Dargush, G.F.: Passive Energy Dissipation Systems in Structural Engineering. Wiley, Chichester (1997)

Soong, T.T., Dargush, G.F.: Passive energy dissipation and active control, Structural Engineering Handbook. CRC Press LLC, Boca Raton (1999)

Soong, T.T., Lai, M.L.: Correlation of experimental results with predictions of viscoelastic damping for a model structure. In: Proceedings of the Damping 91, San Diego, CA (1991)

Tsai, M.H., Chang, K.C.: A study on modal strain energy method for viscoelastically damped structures. WCEE **12** (2000)

Tsai, M.H., Chang, K.C.: A study of the modal strain energy method for viscoelastically damped structures. J Chin. Inst. Eng. **24**(3), 311–320 (2001)

Ungar, E.E., Kerwin, J.R.E.M.: Loss factor of viscoelastic systems in terms of energy concepts. J Am. Acoust. Soc. **34**, 954–957 (1962)

Chapter 4
Modeling of Viscoelastic Dissipative Bracing Systems

Abstract This chapter deals with the theory of viscoelasticity and the discrete models such as, for example, the Kelvin and Maxwell models. The aim of this chapter is to assess the dynamic behavior of viscoelastic dissipative bracing systems taking into account the presence of the brace. In fact, the viscoelastic damper is modeled as the Kelvin model, whose behavior is dependent, in itself, on frequency; the viscoelastic damper-brace component can be studied through the Poynting-Thomson model which presents even more dependence on the frequency. Similarly, the viscous (linear or non-linear) damper-brace component can be studied through the Maxwell model, characterized by a frequency dependent dynamic response. In both cases, because of the frequency dependence, in the dynamic field, dynamic "reduced" magnitudes correspond to the static magnitudes of the viscoelastic dissipative bracing system, in other terms, between the static and dynamic behavior, there is a reduction in the effectiveness of the viscoelastic dissipative bracing system.

4.1 Principles and Models of Linear Viscoelasticity

The theory of linear viscoelasticity is often used because it allows to model and predict the response of many mechanical systems. Applications of the linear viscoelasticity theory are also present in structural engineering with reference to the study of the dynamic response of structural systems equipped with dissipative elements.

Viscoelastic materials are characterized by both elastic (Hooke) and viscous (Newtonian) properties in phase respectively with the displacement and the velocity of motion. These can be considered as two boundary cases of a wide spectrum of behaviors.

Viscoelastic phenomena are, for example, relaxation (stress under constant strain) that occurs when a body is rapidly deformed, and the deformation is maintained constant while the stress induced in the body decreases with time; the

P. Castaldo, *Integrated Seismic Design of Structure and Control Systems*,
Springer Tracts in Mechanical Engineering, DOI: 10.1007/978-3-319-02615-2_4,

creep (strain under constant load) is when a body is rapidly put in tension and the tension is kept constant while the body continues to deform; hysteresis in a body subjected to a cyclic load and occurs when the stress–strain relationship for increasing loads is other than that for decreasing loads.

The more general formulation under the assumption of linearity between the cause and effect is due to Boltzmann (Fung 1965).

In the uni-dimensional case, a simple bar subjected to the force F(t) and elongation u(t) can be considered. The elongation u(t) is caused by the entire load history until time t.

If the function F(t) is continuous and differentiable, in a short interval $d\tau$ at the istant τ the load increasing turns out to be:

$$\left(\frac{dF}{d\tau}\right)d\tau \tag{4.1}$$

At time $t > \tau$, the elongation continues to increase, du(t). In terms of proportionality and dependance on the interval (t – T), the following can be written:

$$du(t) = c(t-\tau)\left(\frac{dF}{d\tau}\right)d\tau \tag{4.2}$$

where c(t) is the function of creep, that is, the elongation response to a unit load variation.

If the origin of time is taken at the beginning of the motion and load, then, summing the contributions of the loading history, the following is obtained:

$$u(t) = \int_0^t c(t-\tau)\left(\frac{dF}{d\tau}\right)d\tau \tag{4.3}$$

Reserving the roles of F and u, the following relationship applies:

$$F(t) = \int_0^t k(t-\tau)\left(\frac{du(\tau)}{d\tau}\right)d\tau \tag{4.4}$$

where k(t) is the relaxation function.

The above equation in the form of differential operators, with reference to the specific terms, can be written as follows (Fung 1965):

$$\sum_{m=0}^{M} a_m \frac{d^m\sigma}{dt^m} = \sum_{n=0}^{N} b_n \frac{d^n\varepsilon}{dt^n} \tag{4.5}$$

where a_m and b_n are constants and the exponents of the derivatives indicate the order of the derivatives, which can be either integer order or fractional derivatives.

The Eq. (4.5) may be, in an equivalent manner, written in integral form (Bagley and Torvik 1983):

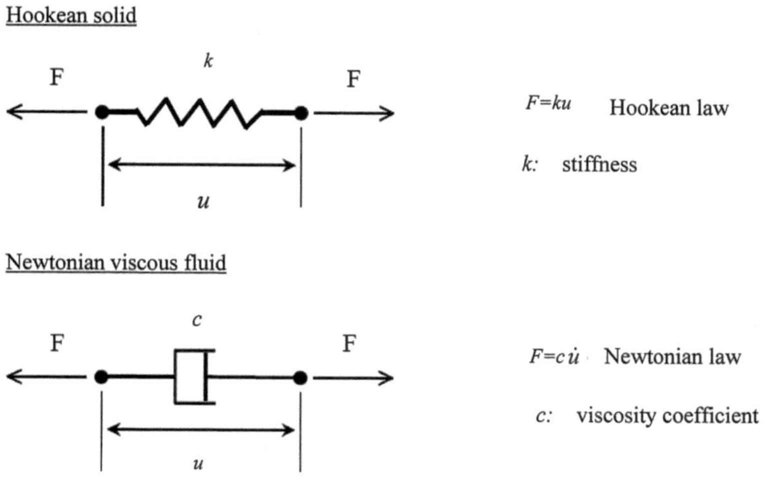

Fig. 4.1 Basic elements

$$\frac{d^{\alpha}f(t)}{dt^{\alpha}} \equiv \frac{1}{\Gamma(1-\alpha)}\frac{d}{dt}\left[\int_{0}^{t}\frac{f(\tau)}{(t-\tau)^{\alpha}}d\tau\right] \tag{4.6}$$

where $0 < \alpha < 1$ and $\Gamma(\cdot)$ is the Gamma function.

The linear viscoelastic behavior, however, is often described by means of mechanical discrete models, composed of the combination of two elements: "linear elastic spring" (Hookean solid) (massless) with spring constant k and "viscous dashpot" (Newtonian Fluid) with viscosity coefficient c (Fig. 4.1).

From the combination of these two basic elements, different models are obtained, widely described in the literature in order to understand the peculiarities of the real mechanical behavior of different materials or different engineering systems. Among them, the most employed models are the Maxwell model, built from the serially connected spring and (linear (Newtonian) or non-linear) dashpot as well as the Voigt-Kelvin model, consisting of the spring and dashpot connected in parallel (Fig. 4.2).

More complex combinations, but no less useful, are for example the models with three elements: the solid Poynting-Thomson model (or solid standard linear SLS) (Fig. 4.3) and solid Jeffreys model (Fig. 4.4).

More general models, having an order greater than those described above, can be constructed by combining several elements (in series or in parallel) and obtaining the N-th order generalized Kelvin model or the N-th order generalized Maxwell model, represented, respectively, in Figs. 4.5 and 4.6.

The Maxwell and Voigt-Kelvin models, with their generalizations, are the discrete theory corresponding to the continuous Boltzman theory in the case of considering the discrete values of the time variable.

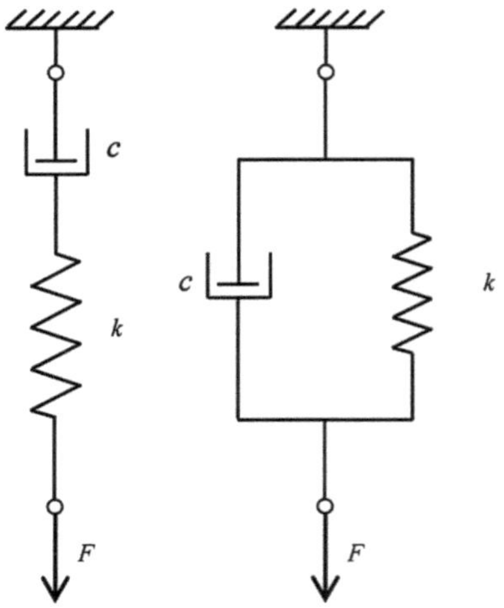

Fig. 4.2 Maxwell and Kelvin-Voigt models

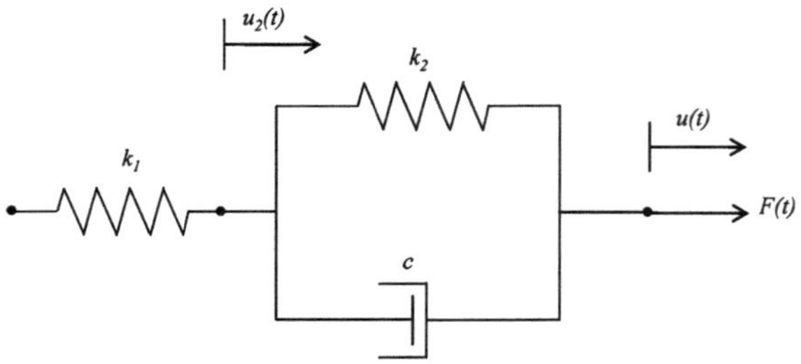

Fig. 4.3 Poynting-Thomson model

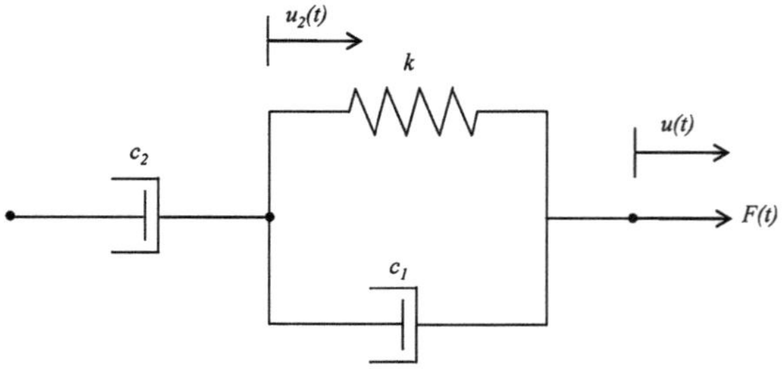

Fig. 4.4 Jeffreys model

The models of the linear visco-elasticity can be studied in the time, frequency and Laplace domain through a mechanical standard model or the fractional derivative model or by means of modified power law representations (Park 2001).

Furthermore, it should be noted that there are numerous models to study the non-linear viscous behavior, whose basic element is a non-linear dashpot (fractional dashpot), with the model being defined as a fractional model. There is, for example, the "fractional Maxwell" model, built from the serially connected spring and "fractional dashpot" and this class of model is studied through an approach to the fractional derivatives.

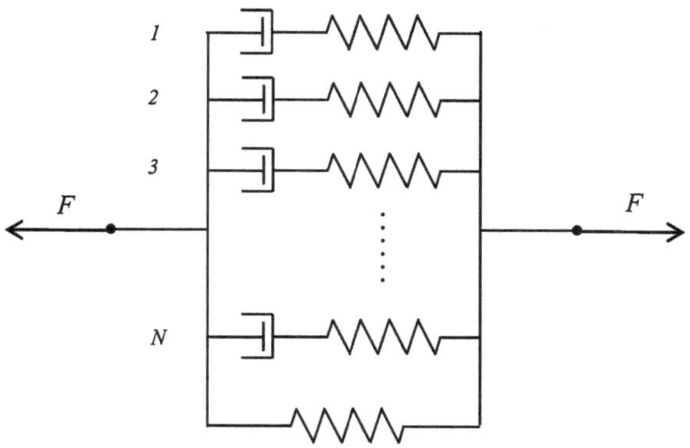

Fig. 4.5 N-th order generalized Kelvin model

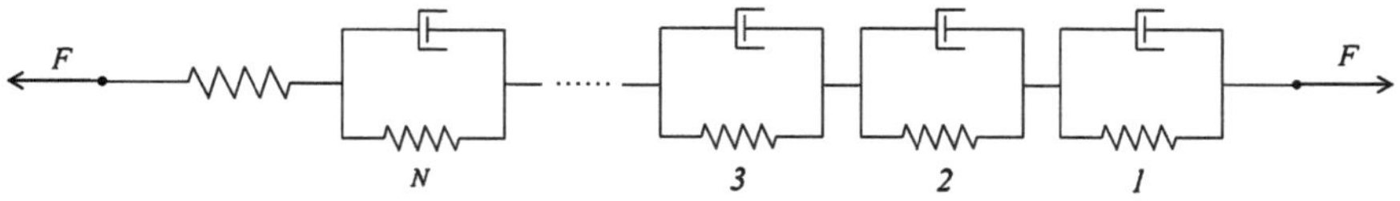

Fig. 4.6 N-th order generalized Maxwell model

4.2 Modeling of the Viscoelastic Damper-Brace Component

A viscoelastic dissipative bracing system defined as a damper–brace component consists of a damper connecting with a brace in series. The mathematical models of such viscous and viscoelastic damper–brace components are discussed in the following sections.

For the analysis of the viscoelastic damper-brace component, the Poynting-Thomson solid standard linear (SLS) model, consisting of a spring connecting with a Kelvin model in series (Fig. 4.7) is used.

In a first step, it is necessary to analyse the Kelvin model representative of the behavior only of the viscoelastic damper. The Kelvin model consists of a spring

Fig. 4.7 Viscoelastic damper-brace component

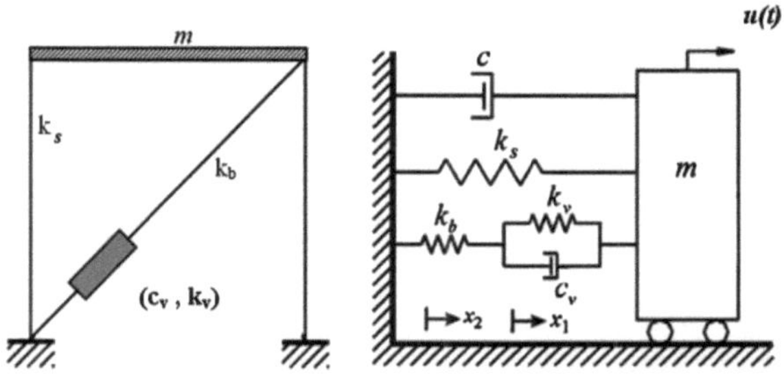

connected with a viscous damper in parallel, and the resistance force of the viscoelastic damper turns out to be:

$$F_v(t) = k_v u_v(t) + c_v \dot{u}_v(t) \tag{4.7}$$

Consider the viscoelastic damper under an imposed sinusoidal displacement time history characterised by a frequency ω_D:

$$u_v(t) = u_0 \sin \omega_D t \tag{4.8}$$

the corresponding function of shear deformation $\gamma(t)$ characterised by a frequency ω_D turns out to be:

$$\gamma(t) = \frac{u_0}{h} \sin \omega_D t = \gamma_0 \sin \omega_D t \tag{4.9}$$

where h is the thickness of the viscoelastic device.

The resistance force of the Kelvin model, with reference to the Eq. (4.7), is:

$$\begin{aligned} F_v(t) &= AG'(\omega_{ve})\gamma(t) + A\frac{G''(\omega_{ve})}{\omega_{ve}}\dot{\gamma}(t) \\ &= AG'(\omega_{ve})\frac{u_0}{h}\sin \omega_D(t) + A\frac{G''(\omega_{ve})}{\omega_{ve}}\frac{u_0}{h}\omega_D \cos \omega_D(t) \end{aligned} \tag{4.10}$$

where A is the cross-sectional area of the device, G′ and G″ are respectively the shear storage modulus and the shear loss modulus. These two moduli describe the dynamic shear behavior of the material and depend on the frequency ω_{ve} of the viscoelastic device, as introduced in Chap. 2. The first term, which is the in-phase portion, represents the elastic modulus that gives a measure of the energy stored and recovered per cycle, and the second term, which is the out-of-phase portion, represents the energy dissipation component that gives a measure of the energy dissipated per cycle.

Consider, from here onwards, $\omega_{ve} = \omega_D$ since the response of the damper is characterized by a dynamic pattern having energy concentrated on its frequency.

The quantity $G''(\omega_{ve})/\omega_{ve}$ is the damping ratio of the device material.

Defining shear storage modulus, shear loss modulus and loss factor, respectively, as follows:

$$k_v(\omega_{ve}) = \frac{AG'(\omega_{ve})}{h} \tag{4.11}$$

$$c_v(\omega_{ve}) = \frac{AG''(\omega_{ve})}{\omega_{ve} h} \tag{4.12}$$

$$\eta(\omega_{ve}) = \frac{G''(\omega_{ve})}{G'(\omega_{ve})} \tag{4.13}$$

and exploiting the Eqs. (4.11)–(4.13), the shear storage modulus can be written as:

$$k_v(\omega_{ve}) = \frac{AG'(\omega_{ve})}{h} = \frac{G'(\omega_{ve})}{G''(\omega_{ve})}\omega_{ve}c_v(\omega_{ve}) = \frac{\omega_{ve}}{\eta(\omega_{ve})}c_v(\omega_{ve}) \quad (4.14)$$

The resistance force of the damper using the Eqs. (4.11) and (4.12) is then equal to:

$$F_v(t) = k_v(\omega_{ve})u_v(t) + c_v(\omega_{ve})\dot{u}_v(t) \quad (4.15)$$

in which the stiffness and viscosity coefficient of the viscoelastic material depend on the frequency ω_{ve} of the viscoelastic device.

In general, the shear deformation $\gamma(t)$ and shear stress $\tau(t)$, assessed as the force divided by the area A, oscillate with the same frequency, but are out of phase since the shear stress also depends on the viscous term correlated to the velocity of relative deformation, as expressed by the following relations:

$$\gamma(t) = \gamma_0 \sin \omega_D t \quad (4.16)$$

$$\tau(t) = \tau_0 \sin(\omega_D t + \delta) \quad (4.17)$$

where, as shown in Fig. 4.8, γ_0 and τ_0 are respectively the shear peak deformation and shear peak stress, and δ is the phase angle, which is equal to $\delta = \omega_{ve}t$. It follows that the time lag can be evaluated as δ/ω_{ve}.

For a given γ_0, both τ_0 and δ depend on ω_{ve}.

When $\omega_{ev}t = \pi/2$, by using the Eqs. (4.10) and (4.17), the following relationship applies:

$$\begin{aligned} \tau(t) &= \tau_0 \sin(\pi/2 + \delta) \\ &= \tau_0 \cos\delta = \gamma_0 G'(\omega_{ve}) \sin(\pi/2) \Rightarrow \frac{\tau_0}{\gamma_0}\cos\delta = G'(\omega_{ve}) \end{aligned} \quad (4.18)$$

When $\omega_{ev}t = 0$, by considering $\omega_{ev} = \omega_D$ and using the Eqs. (4.10) and (4.17), it is obtained:

$$\tau(t) = \tau_0 \sin(\delta) = \gamma_0 G''(\omega_{ve}) \cos(\pi/2) \Rightarrow \frac{\tau_0}{\gamma_0}\sin\delta = G''(\omega_{ve}) \quad (4.19)$$

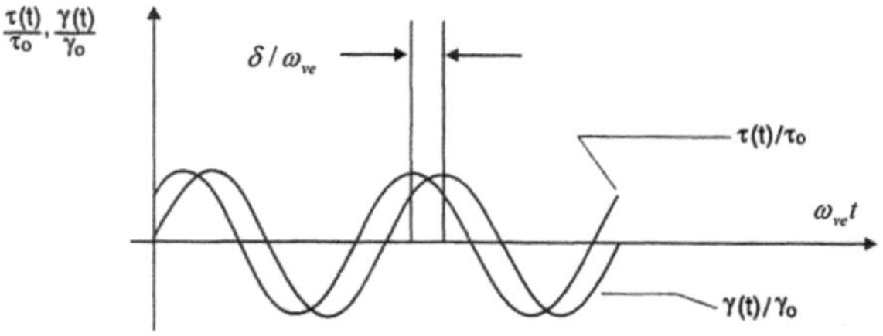

Fig. 4.8 Stress and strain under sinusoidal load

It follows that the shear stress can be written, similarly, by using the Eqs. (4.18) and (4.19), as:

$$\tau(t) = \gamma_0[G'(\omega_D)\sin\omega_D t + G''(\omega_D)\cos\omega_D t] \quad (4.20)$$

The quantity $\gamma_0 \sin\omega_D t$ may be replaced by $\gamma(t)$ and it is obtained:

$$\cos\omega_D t = \frac{1}{\gamma_0 G''(\omega_{ve})}[\tau(t) - G'(\omega_{ve})\gamma(t)] \quad (4.21)$$

By exploiting the Eq. (4.20) and using the identity $\sin^2\omega_D t + \cos^2\omega_D t = 1$, the following tension-deformation relationship is obtained:

$$\tau(t) = G'(\omega_{ve})\gamma(t) \pm G''(\omega_{ve})\left[\gamma_0^2 - \gamma^2(t)\right]^{1/2} \quad (4.22)$$

that defines an ellipse shown in Fig. 4.9, whose area provides the energy dissipated by the viscoelastic material per unit volume and per cycle of vibration. It is given by:

$$E_H = \int_0^{\frac{2\pi}{\omega}} \tau(t)\dot{\gamma}(t)dt = \pi\gamma_0^2 G''(\omega_{ve}) \quad (4.23)$$

As it can be noted from the Eq. (4.22), the first term of the shear stress is the portion in phase and the second term is the part that is the out of phase component of the energy dissipated. This is most visible when rewriting the Eq. (4.20) in the form:

$$\tau(t) = G'(\omega_{ve})\gamma(t) + \frac{G''(\omega_{ve})}{\omega_{ve}}\dot{\gamma}(t) \quad (4.24)$$

The equivalent damping ratio or rather the damping value that would characterize an equivalent viscous device to dissipate the same energy is equal to:

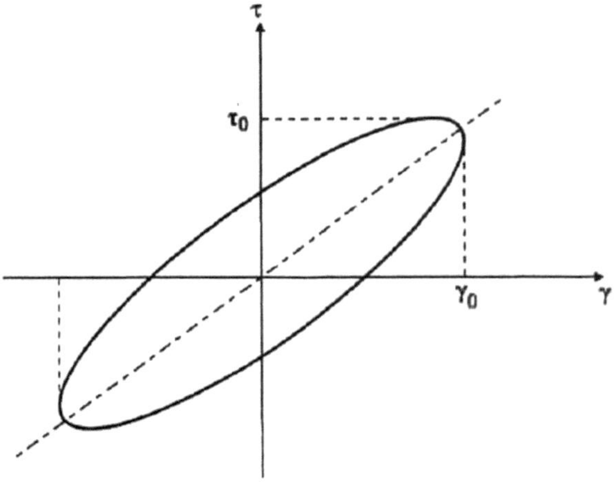

Fig. 4.9 Plot of stress versus strain

$$
\begin{aligned}
\xi &= \frac{c_v(\omega_{ve})}{2\omega m} = \frac{G''(\omega_{ve})}{\omega_{ve}}\frac{A}{h}\frac{1}{2\omega_{ve}m} = \frac{G''(\omega_{ve})}{\omega_{ve}}\frac{A}{h}\frac{1}{2\sqrt{k_v(\omega_{ve})m}} \\
&= \frac{G''(\omega_{ve})}{\omega_{ve}}\frac{A}{h}\frac{1}{2\sqrt{\frac{AG'(\omega_{ve})}{h}m}} \\
&= \frac{G''(\omega_{ve})}{\omega_{ve}}\frac{A}{h}\frac{1}{2\sqrt{\frac{AG'(\omega_{ve})}{h}m}}\frac{\sqrt{G'(\omega_{ve})}}{\sqrt{G'(\omega_{ve})}} \\
&= \frac{G''(\omega_{ve})}{2G'(\omega_{ve})\omega_{ve}}\sqrt{\frac{AG'(\omega_{ve})}{hm}} = \frac{G''(\omega_{ve})}{2G'(\omega_{ve})\omega_{ve}}\sqrt{\frac{k_v(\omega_{ve})}{m}} \\
&= \frac{G''(\omega_{ve})}{\omega_{ve}}\left(\frac{\omega_{ve}}{2G'(\omega_{ve})}\right) = \frac{G''(\omega_{ve})}{2G'(\omega_{ve})}
\end{aligned}
\tag{4.25}
$$

By exploiting the relationship (4.13), the damping ratio becomes:

$$
\xi = \frac{\eta}{2} \tag{4.26}
$$

Recall that $G'(\omega_D)$ is defined *elastic modulus* of the viscoelastic material and is a measure of the energy stored and recovered per cycle, and that $G''(\omega_D)$ is called a *dissipation modulus* and provides a measure of the dissipated energy per cycle, the loss factor η can also be defined, by using the Eqs. (4.18) and (4.19), as:

$$
\eta = \frac{G''(\omega_{ve})}{G'(\omega_{ve})} = \tan\delta \tag{4.27}
$$

that is often used as a measure of the capacity to dissipate energy of the viscoelastic material as expressed in the Eq. (4.26).

In order to model the viscoelastic damper connecting with an elastic brace in series and to assess the dependence on the frequency of the coefficients of Poynting-Thomson model, which will be more marked than that analyzed on loss factor of the single device (Soong and Constantinou 1994; Soong and Dargush 1997), it is possible in this model, since it is a series, correlating the relative displacement and the relative velocity between the ends of the viscoelastic device, respectively, to the relative displacement and relative velocity between the ends of the "viscoelastic damper-brace component" and, therefore, to obtain the differential equation that governs the dynamic response of such a system (Poynting-Thomson model), in the time domain, by following an approach of mechanical standard model (Park 2001). This equation turns out to be:

$$
F_{vb}(t) + \frac{c_v}{k_b + k_v}\dot{F}_{vb}(t) = k_{vb}\left[u_{vb}(t) + \frac{c_v}{k_v}\dot{u}_{vb}(t)\right] \tag{4.28}
$$

(Note that, for simplicity of writing, the dependence of c_v and k_v on ω_{ve} is implied) where k_{vb} is the stiffness of the viscoelastic damper-brace component or rather the series $k_{vb} = \frac{k_v k_b}{k_v + k_b}$ of the two stiffnesses, c_v is the viscosity coefficient of the viscoelastic device, k_b is the axial stiffness of the brace, k_v is the stiffness of the viscoelastic device, $F_{vb}(t)$ is the resistance force of the viscoelastic damper-brace component, $\dot{F}_{vb}(t)$ is its derivative with respect to time t, $u_{vb}(t)$ is the relative displacement between the ends of the viscoelastic damper-brace component and $\dot{u}_{vb}(t)$ its time derivative.

Consider a (damped) harmonic displacement time history with frequency ω_D and amplitude u_0, unless the transition regime represented by the solution of the homogeneous equation associated, the resistance force of the viscoelastic damper-brace component, which is the particular solution and representative of the condition in stationary regime, can be expressed as:

$$\begin{aligned} F_{vb}(t) &= \left[k_{vb} + \tilde{\tau}\omega_D \left(\frac{k_{vb} c_v \omega_D - \tilde{\tau}\omega_D k_v k_{vb}}{(1+\tilde{\tau}^2\omega_D^2)k_v}\right)\right] u_{vb}(t) + \left[\frac{k_{vb} c_v - \tilde{\tau} k_v k_{vb}}{(1+\tilde{\tau}^2\omega_D^2)k_v}\right] \dot{u}_{vb}(t) \\ &= k'(\omega_D)[u_0 \sin(\omega_D t)] + c'(\omega_D)\omega_D[u_0 \cos(\omega_D t)] \end{aligned} \tag{4.29}$$

where $\tilde{\tau} = \frac{c_v}{k_v + k_b}$, $k'(\omega_D)$ and $c'(\omega_D)$ are frequency dependent dynamic moduli of the viscoelastic damper-brace component, also referred to as, respectively, "apparent stiffness" and "apparent damping". After simple mathematical operations, the latter are defined by the following relations (4.30) and (4.31):

$$k_{vb}(\omega_D) = k'(\omega_D) = \frac{k_b k_v^2 + k_b^2 k_v + k_b c_v^2 \omega_D^2}{(k_b + k_v)^2 + c_v^2 \omega_D^2} \tag{4.30}$$

$$c_{vb}(\omega_D) = c'(\omega_D) = \frac{k_b^2 c_v}{(k_b + k_v)^2 + c_v^2 \omega_D^2} \tag{4.31}$$

In the frequency domain through the Fourier transform and by using the complex theory of linear viscoelasticity (Clough and Penzien 1993; Ou et al. 2007), the terms, defined by the Eqs. (4.30) and (4.31), can be interpreted as storage and loss moduli. In Eqs. (4.32) and (4.33), the "storage modulus" and "loss modulus" are defined.

$$k'_{vb}(\omega_D) = k'(\omega_D) = \frac{k_b k_v^2 + k_b^2 k_v + k_b c_v^2 \omega_D^2}{(k_b + k_v)^2 + c_v^2 \omega_D^2} \tag{4.32}$$

$$k''_{vb}(\omega_D) = c'(\omega_D) = \frac{k_b^2 c_v \omega_D}{(k_b + k_v)^2 + c_v^2 \omega_D^2} \tag{4.33}$$

In the frequency domain, the loss factor of the viscoelastic system is expressed as:

$$\eta_{vb}(\omega_D) = \frac{k''_{vb}(\omega_D)}{k'_{vb}(\omega_D)} = \frac{\tau\omega_D}{\frac{k_v^2}{k_b^2} + \frac{k_v}{k_b} + \tau^2\omega_D^2} \tag{4.34}$$

where $\tau = \frac{c_v}{k_b}$ is known as the relaxation time.

Similarly, by defining $k'_v(\omega_D) = k_v$ and $k''_v(\omega_D) = c_v\omega_D$ the loss factor of the viscoelastic damper is:

$$\eta_v(\omega_D) = \frac{k''_v(\omega_D)}{k'_v(\omega_D)} \tag{4.35}$$

The damping and stiffness reduction coefficients are, in conclusion, defined as:

$$\beta_{vc} = \frac{k''_{vb}(\omega_D)}{k''_v(\omega_D)} = \frac{1}{\left(1 + \frac{1}{\eta_v\eta_{vb}}\right)^2 + \frac{1}{\eta_{vb}{}^2}} \tag{4.36}$$

$$\beta_{vk} = \frac{k'_{vb}(\omega_D)}{k'_v(\omega_D)} = \frac{\eta_{vb}\left(\frac{1}{\eta_v} + \eta_v + \eta_{vb}\right)}{\eta_{vb}^2 + 2\eta_{vb}\frac{1}{\eta_v} + \frac{1}{\eta_v^2} + 1} \tag{4.37}$$

From the last equations, it is possible to highlight that the stiffness of the brace connecting the viscoelastic damper to the structure actually affects the damper efficiency significantly, which depends both on the viscoelastic damper-brace component parameters and the natural frequencies of the structure. In most cases, its influence on the performance of the damper should not be neglected; in other words, the brace stiffness should not be approximately treated as infinite in the seismic response analysis of the structure with dampers (Ou et al. 2007).

4.3 Modeling of the (Linear) Viscous Damper-Brace Component

The resistance force of the linear viscous damper, in compliance with the Eq. (2.28), is equal to:

$$F_d(t) = c\dot{u}_d(t) \tag{4.38}$$

where $\dot{u}_d(t)$ is the relative velocity between the ends of the device.

In order to model the viscous damper connecting with an elastic brace in series or rather to study the dependence on the frequency of the viscous damper-brace component, the Maxwell model, consisting of a spring connecting with a linear (Newtonian) damper in series (Figs. 4.10 and 4.11) is used.

It is also possible in the Maxwell model, since it is a series, correlating the relative displacement and the relative velocity between the ends of the viscous

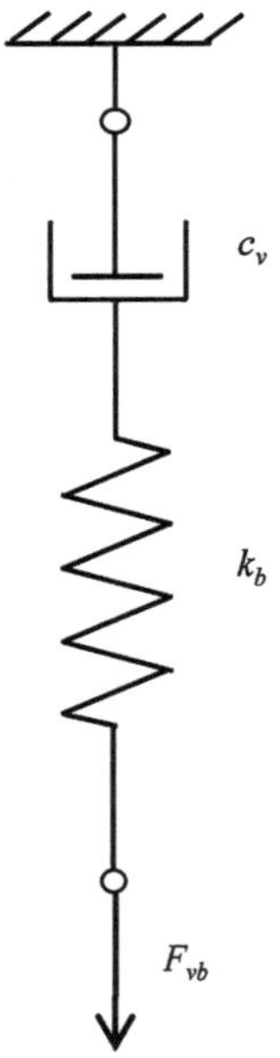

Fig. 4.10 Maxwell model

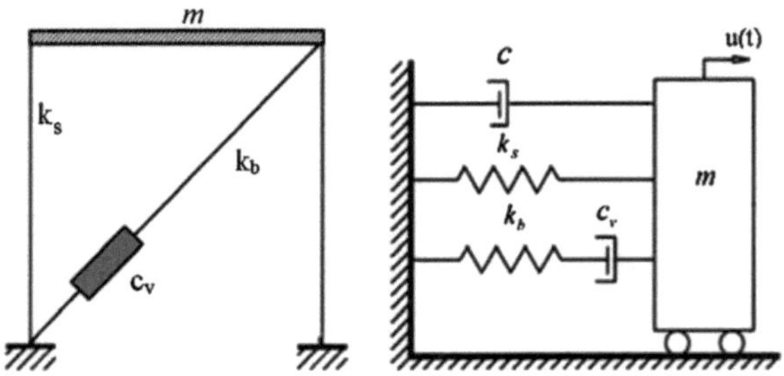

Fig. 4.11 Viscous damper-brace component

device, respectively, to the relative displacement and relative velocity between the ends of the viscous damper-brace component and, then, to obtain the differential equation that governs the dynamic response of the serially connected brace and linear viscous damper (Maxwell model), in the time domain, by following an approach of mechanical standard model (Park 2001), that is:

$$F_{vb}(t) + \frac{c_v}{k_b}\dot{F}_{vb}(t) = c_v \dot{u}_{vb}(t) \tag{4.39}$$

where c_v is the viscosity coefficient of the viscous fluid device, k_b is the axial stiffness of the brace, the ratio $\tau = c_v/k_b$ is defined relaxation time, $F_{vb}(t)$ is the resistance force of the viscous damper-brace component (Maxwell model), $\dot{F}_{vb}(t)$ is its derivative with respect to time t, $u_{vb}(t)$ is the relative displacement between the ends of the viscous damper-brace component (Maxwell model) and $\dot{u}_{vb}(t)$ its time derivative.

Consider a (damped) harmonic displacement time history with frequency ω_D and amplitude u_0, unless the transition regime represented by the solution of the homogeneous equation associated, the resistance force of the viscous damper-brace component (Maxwell model), which is the particular solution and representative of the condition in stationary regime, can be expressed as:

$$\begin{aligned} F_{vb}(t) &= k_b \frac{\tau^2 \omega_D^2}{1+\tau^2\omega_D^2} u_{vb}(t) + c_v \frac{1}{1+\tau^2\omega_D^2} \dot{u}_{vb}(t) \\ &= k'(\omega_D)[u_0 \sin(\omega_D t)] + c'(\omega_D)\omega_D[u_0 \cos(\omega_D t)] \end{aligned} \tag{4.40}$$

in which the frequency dependent dynamic moduli $k'(\omega_D)$ and $c'(\omega_D)$ of the viscous damper-brace component (Maxwell model), also called, respectively, "apparent stiffness" and "apparent damping", are defined by the following relations (4.41) and (4.42):

$$k_{vb}(\omega_D) = k'(\omega_D) = k_b \frac{\tau^2\omega_D^2}{1+\tau^2\omega_D^2} \tag{4.41}$$

$$c_{vb}(\omega_D) = c'(\omega_D) = c_v \frac{1}{1+\tau^2\omega_D^2} \tag{4.42}$$

In the frequency domain through the Fourier transform and by using the complex theory of linear viscoelasticity (Clough and Penzien 1993; Ou et al. 2007), the terms, defined by the Eqs. (4.41) and (4.42), can be interpreted as storage and loss moduli. In Eqs. (4.43) and (4.44), the "storage modulus" and "loss modulus" are respectively defined.

$$k'_{vb}(\omega_D) = k_b \frac{\tau^2\omega_D^2}{1+\tau^2\omega_D^2} \tag{4.43}$$

$$k''_{vb}(\omega_D) = c_v \frac{1}{1+\tau^2\omega_D^2} \omega_D \tag{4.44}$$

In the frequency domain, the loss factor is expressed as:

$$\eta_{vb}(\omega_D) = \frac{k''_{vb}(\omega_D)}{k'_{vb}(\omega_D)} = \frac{k_b}{c_v\omega_D} = \frac{1}{\tau\omega_D} \tag{4.45}$$

The damping and stiffness reduction coefficients are, therefore, defined as:

$$\beta_{vc}(\omega_D) = \frac{c_{vb}(\omega_D)}{c_v} = \frac{c'(\omega_D)}{c_v} \tag{4.46}$$

$$\beta_{vk}(\omega_D) = \frac{k_{vb}(\omega_D)}{k_b} = \frac{k'(\omega_D)}{k_b} \tag{4.47}$$

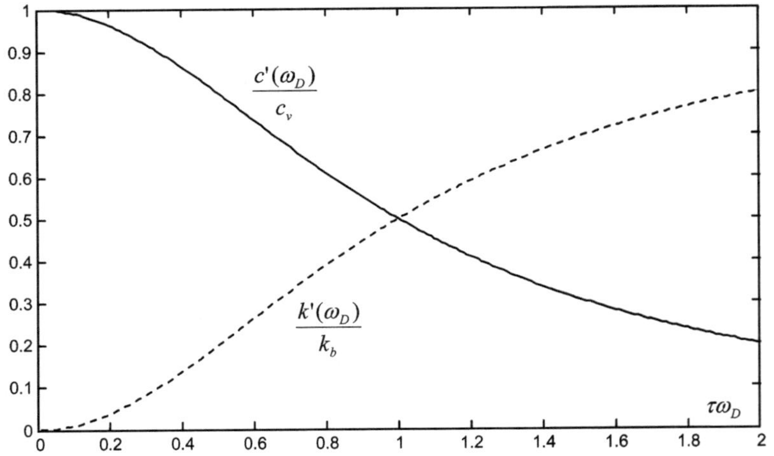

Fig. 4.12 The damping and stiffness reduction coefficients for different values of the product $\tau\omega_D$

The variation of the damping and stiffness reduction coefficients of the viscous damper-brace component with the product $\tau\omega_D$ is illustrated in Fig. 4.12.

By evaluating the limits at zero and infinity, the values of the two coefficients, respectively, are:

$$\lim_{\tau\omega\to 0} \frac{c'(\omega_D)}{c_v} = 1 \tag{4.48}$$

$$\lim_{\tau\omega\to +\infty} \frac{c'(\omega_D)}{c_v} = 0 \tag{4.49}$$

$$\lim_{\tau\omega\to 0} \frac{k'(\omega_D)}{k_b} = 0 \tag{4.50}$$

$$\lim_{\tau\omega\to +\infty} \frac{k'(\omega_D)}{k_b} = 1 \tag{4.51}$$

Similar to the case of viscoelastic dampers, it is possible to highlight that the stiffness of the brace connecting the viscous damper to the structure actually affects the damper efficiency significantly, which depends both on the viscous damper-brace component parameters and the natural frequencies of the structure. In most cases, its influence on the performance of the damper should not be neglected; in other words, the brace stiffness should not be approximately treated as infinite in the seismic response analysis of the structure with dampers (Ou et al. 2007).

Within the framework of analysis developed and illustrated in the following chapters, the Maxwell model is employed and the symbols c_v and k_b are used to define, respectively, the static viscosity coefficient and brace stiffness and the symbols c' and k' are used to define, respectively, the dynamic viscosity coefficient and brace stiffness.

References

Bagley R.L., Torvik, P.J.: A theoretical basis for the application of fractional calculus to viscoelasticità. J. Rheol. **27**, 201–210 (1983)
Clough R.W., Penzien J.: Dynamics of Structures, 2nd edn. McGraw-Hill, New York (1993)
Fung, Y.C.: Foundations of Solid Mechanics. Prentice-Hall, New Jersey (1965)
Ou, J.P., Long, X., Li, Q.S.: Seismic response analysis of structures with velocity-dependent dampers. J. Constr. Steel Res. **63**(5), 628–638 (2007)
Park, S.W.: Analytical modelling of viscoelastic dampers for structural and vibration control. Int. J. Solids Struct. **38**(44), 8065–8092 (2001)
Soong, T.T., Constantinou, M.C.: Passive and Active Structural Vibration Control in Civil Engineering. Springer, New York (1994)
Soong, T.T., Dargush, G.F.: Passive energy dissipation systems in structural engineering. Wiley and Sons Ltd, Chichester (1997)

Chapter 5
Integrated Design of Proportionally Damped Framed Structural Systems Equipped with Viscous Devices

Abstract This chapter deals with the integrated design of the elastic structural system and the viscoelastic dissipative bracing system to achieve an expected seismic design displacement. The variables that characterize the design problem, their domain and the steps of the proposed methodology are illustrated and commented in details by taking into account the concepts described in the previous chapters. A set of seven historical unscaled acceleration records is selected to develop dynamic analyses by testing the proposed integrated design methodology. With reference to an equivalent SDOF integrated system, a cost index, assumed as an optimized objective function and defined on the design variables, is described in order to find the optimal design in economic terms, or rather, the economically optimal combination of the design variables for each expected seismic target performance. Finally, the extension of the results, developed on the substitute structure, to a viscoelastically and proportionally damped MDOF framed integrated system is explained in order to design the least expensive regular system.

5.1 Integrated Seismic Design Philosophy

Over the past years, structure and control systems have been independently designed and, in some cases, optimized. Over the last 30 years, a lot of research has been developed on the integrated optimal design of structural/control systems.

Specifically, the idea is to consider both the viscoelastic resources of the bracing-damper system as well as the elastic ones of the structural system from the beginning of design in order to obtain an integrated design of elastic structural/viscoelastic control systems. In this sense, the main variables that characterize the dynamic response of both systems are investigated as variables to be simultaneously optimized within the same design, the so-called "integrated design" (Ramana 1990; Smith et al. 1992; Zhang and Soong 1992; Cimellaro et al. 2009; Soong and Cimellaro 2009; Rama et al. 2013; Xu 2013; Xu et al. 2012) (Fig. 5.1).

P. Castaldo, *Integrated Seismic Design of Structure and Control Systems*,
Springer Tracts in Mechanical Engineering, DOI: 10.1007/978-3-319-02615-2_5,

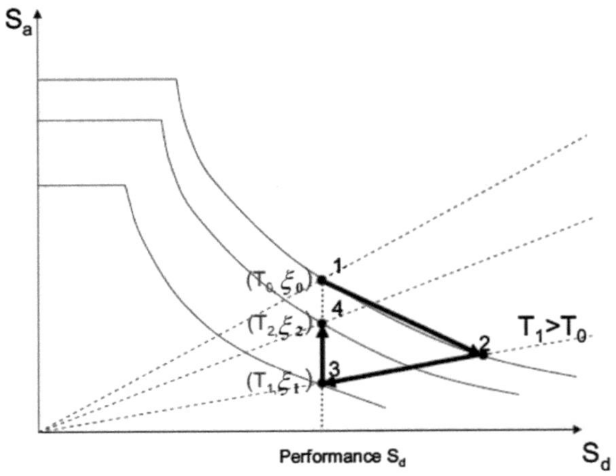

Fig. 5.1 Redesign procedure in S_a–S_d plan (modified from Cimellaro et al. 2009)

By following this new design philosophy, the study proposes a simplified method, based on a performance approach, aimed at achieving an integrated optimal design of elastic structural/vicoselastic control systems minimizing a cost index, assumed as an optimized objective function. In other words, the possibility of assigning the vibration control will be explored, with it traditionally being only assigned to the lateral elastic stiffness of the system as well as the viscoelastic ones of a viscoelastic dissipative bracing system.

Assuming that the dynamic behavior of a n-degrees-of-freedom framed system is governed by the first mode, the methodology of integrated design is conducted on an equivalent single-degree-of-freedom structural system. In this context, the lateral stiffness of the structure, with it being representative of its elastic dynamic response, the stiffness of the brace and viscosity coefficient of the viscous damper, connected with brace in series have been regarded as design variables. The last two variables identify the non-linear dynamic response of the viscous damper-brace component, whose effect on the dynamic response of the integrated system has been explicitly considered.

The search for the optimal combination of design variables is performed through the dynamic analyses on a substitute (SDOF) integrated system (Shibata and Sozen 1976; Priestley et al. 1996) by considering a set of seven historical unscaled acceleration records. These recordings have been chosen in such a way that they are compatible (CEN 2005) in average with the elastic spectrum ($\xi = 5$ %) relative to the ultimate limit state ULS, life safe, provided by the new Italian seismic code NTC08 (NTC 2008) for a specific site in Italy.

The optimal design, related to the equivalent SDOF integrated system corresponding to expected performances, is represented by the design variables, valued on the average seismic displacement demand related to the acceleration records, that minimize a cost index of the integrated system.

5.2 Displacement-Based Integrated Seismic Design

The performance-based design is the new seismic design philosophy of the new codes, in which the performance levels depend on the importance and function of the structure. In fact, in a structure with a high level of importance and function, the damage level must be negligible after earthquake events characterized by high return period T_R, i.e. high seismic intensity. A summary of the new design philosophy called performance-based design is illustrated in the so-called performance design objective matrix (Fig. 5.2), in which it is possible to arrange the performance level to be achieved by a structure (also known as seismic limit state) in function of the different levels of expected seismic intensity.

Arranging the performance is the main step of the design, the whole process, in fact, should be based on the achievement of the objectives. These objectives are in general related to the limit values of the measurable parameters of the structural response, such as interstory drift, the ductility demand, the dissipated energy, or more in general, damage indeces that represent the combined effect of these factors. Once the performance objective has been indicated, satisfying these limit values becomes the criterion of verification to be used. The identification of these limits is still being discussed, with there being different opinions about the definition of indices that best represent damage by considering the overall behavior of a structure.

The evaluation of the seismic response of a structure may be conducted by means of a linear/non-linear static/dynamic analysis. All four methods are based on displacement or variables directly or indirectly linked to it in order to carry out assessments of control such as procedures presented in FEMA 273 and 274 (ATC 1997a, b). This is a considerable change in the practice of seismic design since the focus of the analysis, design and evaluation has moved from force to displacement. The actions on the elements of a building are classified as either displacement-control

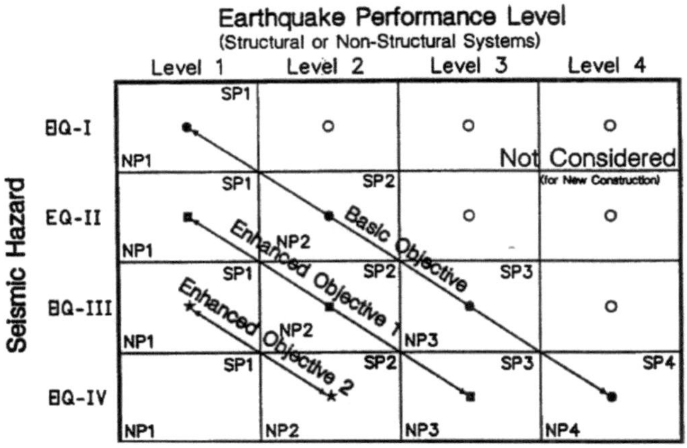

Fig. 5.2 Performance design objective matrix (PDOM)

actions for ductile mechanisms such as bending in the beams, or as force control actions for brittle mechanisms as the shear in columns. The limits of ultimate rotation capacity for displacement-control actions were presented in the chapters about the materials of FEMA 273 (ATC 1997a) to check the estimation of the rotation capacity using displacement-based methods.

The (direct) displacement-based seismic design DBD (Priestley et al. 2007), which considers the displacement as a key design parameter, is considered an effective method for the implementation of performance-based seismic design by using the deformation capacity and ductility. In fact, there are various contributions in literature relating to the development of more or less simplified performance-based design procedures of structural systems equipped with passive dissipation devices by using the capacity spectrum method or the response spectrum in terms of acceleration-displacement (ADRS) or by using the response spectrum in terms of displacement and the capacity curve of the structure (Sullivan et al. 2003; Kim et al. 2003; Kim and Seo 2004; Lin et al. 2003; Kim and Choi 2006; ATC-40 (ATC 1996), FEMA-356 (ATC 2000), FEMA-273 (ATC 1997a); Freeman 1998; SEAOC 1999).

In the proposed integrated design methodology, which will be presented in the following paragraphs, concerning linear structural systems equipped with viscous devices, a displacement-based seismic design (DBD) through the use of spectra in terms of displacement has been employed.

For example, the interstory drift limits related to the four performance levels corresponding to specific levels of seismic intensity reported in ATC-40 (ATC 1996) are shown in Table 5.1.

In Table 5.1 V_i and P_i are, respectively, the total shear force and the value of the gravitational (dead and live) loads related to the i-th floor.

Other examples of limitations reported in literature (Ferraioli et al. 2005; Ghobarah 2001) are shown in Table 5.2, in which (IDI: interstory drift index) the damage index is related to the performance levels corresponding to different seismic intensity levels.

Table 5.1 Displacemnet limits in ATC-40 (ATC 1996)

Performance levels				
Interstory drift	Full Operation	Operational	Life safety	Collapse
Maximum total displacement/interstory height	0.01	0.01–0.02	0.02	0.33 V_i/P_i
Maximum inelastic displacement/interstory height	0.005	0.005–0.015	*Not considered*	*Not considered*

Table 5.2 Damage indeces expressed in terms of interstory drift (Ferraioli et al. 2005)

Seismic performance levels	Return period (T_R)	Structural damage		Non structural damage	
		Damage Index	Failure probability	Index IDI	Failure probability
Full Operational	30	0.20	0.20	0.003	0.30
Operational	75	0.40	0.20	0.006	0.30
Life safety	475	0.60	0.10	0.015	0.20
Collapse Prevention	970	0.80	0.10	0.020	0.20

In the philosophy of the above-described displacement-based design (DBD), the proposed integrated design has been developed by considering as the performance parameter of the integrated system, consisting of elastic structural and viscoelastic control systems, the relative displacement between top of the structure and ground, chosen ranging from 1 mm to 10 cm.

5.3 Integrated Design Approach

The design of passive dissipation devices are generally used as a seismic rehabilitation technique of an existing structural system. This approach is fully justified only in the application of the seismic rehabilitation of monumental and historical buildings, where it is not possible, for obvious reasons, to modify the primary structure. This approach, as already discussed, appears inherently limited, since it does not provide for the possibility of an integrated design between the structure and control system which explicitly takes into account the dynamic interaction in terms of performance between the two components of the integrated system. In the literature of structural passive control related to linear viscous dampers, there are very few research studies dealing with the optimal integrated design between these two components of the integrated system that explicitly take into account their dynamic behavior.

Therefore, an integrated design methodology of structural systems with linear viscous devices is proposed so that the dynamic response of the system achieves an expected seismic design displacement. The methodology proposed is aimed, in particular, at the identification of the optimal design, in economic terms, between all the solutions obtainable by an integrated design of equivalent linear single-degree-of-freedom (SDOF) structural systems equipped with linear viscous dampers, which present a dynamic response compatible with an expected target performance.

The dynamic response equation of the SDOF integrated system in the state space form, as already seen, is:

$$\dot{\mathbf{z}}(t) = \mathbf{A}\mathbf{z}(t) + \mathbf{b}\ddot{u}_g(t) \tag{5.1}$$

where

$$\mathbf{z}(t) = \begin{bmatrix} u(t) \\ \dot{u}(t) \end{bmatrix} \quad \mathbf{A} = \begin{bmatrix} 0 & 1 \\ -\omega^2 & -2\xi\omega \end{bmatrix} \quad \mathbf{b} = \begin{bmatrix} 0 \\ 1 \end{bmatrix} \tag{5.2}$$

where u(t) represents both the relative displacement of the structural SDOF system respect to the ground, indicated with x(t) in Chap. 3, that the relative displacement, less than $\cos\vartheta$, between the ends of the viscous damper-brace component, indicated with u_{vb} in Chap. 4; similarly $\dot{u}(t)$ is the relative velocity between both the top of the structural system that the upper end of the viscous damper-brace component and the ground; ω the natural (undamped) frequency of the integrated system, where k is the sum of the lateral stiffness of the main structure, k_s, and the dynamic stiffness of the dissipative bracing system, $k'(\omega_D)$; ξ is the viscous damping ratio seen as the sum (Eq. (5.3)) of the viscous inherent damping ratio of the main structure assumed to be equal $\xi_s = 2\,\%$ and to the contribution ξ_d of the dynamic viscosity coefficient of the dissipative bracing system, $c'(\omega_D)$; ω_D is the damped frequency of the response of the integrated system equipped with viscous damper.

$$\xi = \xi_s + \xi_d \tag{5.3}$$

Therefore, for a given seismic design displacement u and set a natural vibration period T, the variables that influence the response and that, therefore, are considered as the design variables are: the lateral stiffness of the structural system k_s, the static stiffness k_b and the static viscosity coefficient c_v of the linear viscoelastic dissipative bracing system. The dynamic behavior of the viscous damper-brace component is explicitly considered in the analyses. In Fig. 5.3, the analytical model considered in the analysis is shown.

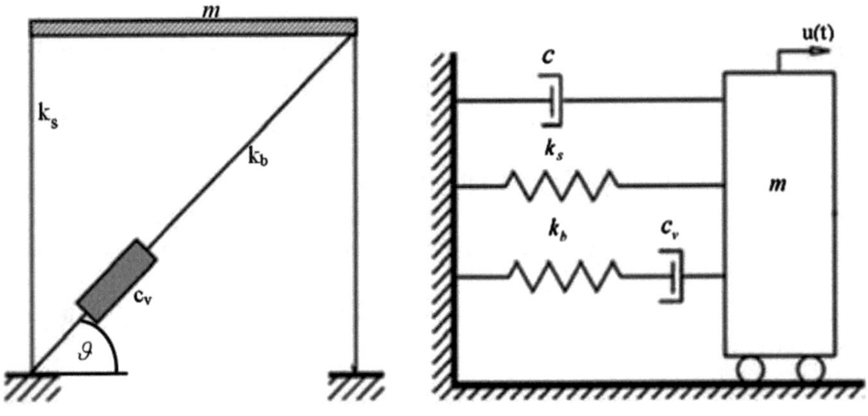

Fig. 5.3 Analytical model of the equivalent SDOF system with viscous damper

The lateral stiffness of the structural system is considered variable between 20 % and 100 % of the total lateral one and, assigned a value to that variable, its complement to 100 is equal to the dynamic stiffness of the passive control system, connected to the structural system in parallel. The value of 20 % was chosen as a minimum value $k_{s,\min}$ necessary for supporting the gravity loads, as indicated in several scientific studies (Cimellaro et al. 2009; Soong and Cimellaro 2009; Symans et al. 2008).

Another constraint has also been considered, in a second phase, on the maximum value of the damping ratio assumed to be equal to 40 %.

The parametric analysis has been carried out considering the natural vibration period of the integrated system ranging from 0 to 3 s thus to obtain a spectral representation of the results. In this way, the integrated design can be carried out by considering the undamped natural vibration period both fixed and as a design variable. The above-described variability may, for example, oscillate in a range whose ends can be represented by the values of the period relating, respectively, to unbraced and braced systems: there are several formulations in literature to estimate the period of a structural system with or without braces depending on the material (Tremblay 2005; Chrysanthakopoulos et al. 2006; Jalali and Milani 2004; Crowley and Pinho 2009). It follows that the procedure is applicable both to the case of existing structural systems and new constructions.

In this perspective, the performance in terms of relative displacement respect to the ground has been chosen as variable in a wide range from 1 mm up to 10 cm.

The wide variation of the period is justified by the possibility that the single-degree-of-freedom (SDOF) system may be representative of the dynamic behavior of planar multi-degrees-of-freedom (MDOF) systems characterised by regularity in elevation and, therefore, having a dynamic behavior governed by the first mode (Shibata and Sozen 1976; Priestley et al. 1996). In this case, the aim is to have the optimal design values of stiffness and viscosity and their relationship to be extended, then, in elevation to each structural level of the corresponding viscoelastically and proportionally damped multi-degrees-of-freedom (MDOF) framed integrated system.

The integrated design can be, therefore, considered as a bound constrained problem (Fig. 5.4), whose optimal design turns out to be the combination of the three design variables which minimizes a cost index of the integrated system. Then, for given target performance, period of the equivalent single-degree-of-freedom integrated system and for fixed relative ratios of cost, the economically optimal values of the viscosity coefficient and stiffness to assign to the viscoelastic dissipative bracing system corresponding to an economically optimal value of the lateral stiffness of the structural system are obtained. In a second phase, by considering the period of the integrated system as a design variable and imposing a constraint on the maximum value of the damping ratio, it is possible to obtain the optimal values of the design variables considered, too.

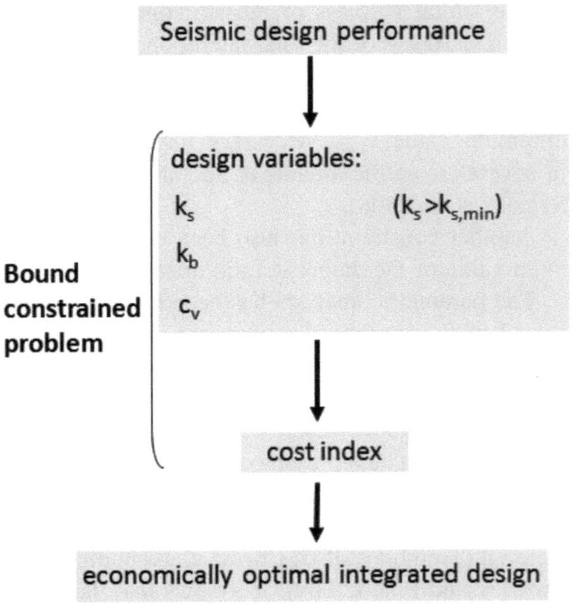

Fig. 5.4 Integrated design: a bound constrained problem

5.4 Integrated Design Methodology

The methodology followed in order to research the optimal integrated design is illustrated in Fig. 5.5 by means of a flow chart.

The first step consists of the evaluation of the viscous damping ratio ξ demanded by each selected seismic registration (selected in Sect. 5.4.1), depending on the seismic performance and the natural vibration period T of the single-degree-of-freedom integrated system in order to obtain the displacement demand spectra for different values of damping ratio with different seismic design displacements (such spectra have been object of several contributions in the literature, as illustrated in Miranda and Bertero 1994). These analyses are carried out in the state space by implementing the Eqs. (5.1)–(5.2). It follows that it is possible to evaluate the average displacement spectrum by averaging the spectra in terms of displacement of the selected acceleration records set (CEN 2005). The evaluation of the average displacement spectrum allows to estimate the design damping ratio for each period and average design performance, which is the first element to be known through the proposed methodology.

Having evaluated ξ for every considered average displacement and frequency ω (period T), the second step consists of estimating the frequency ω_D (period T_D) of the damped response through the following relation:

$$\omega_D = \omega\sqrt{1 - \xi^2} \tag{5.4}$$

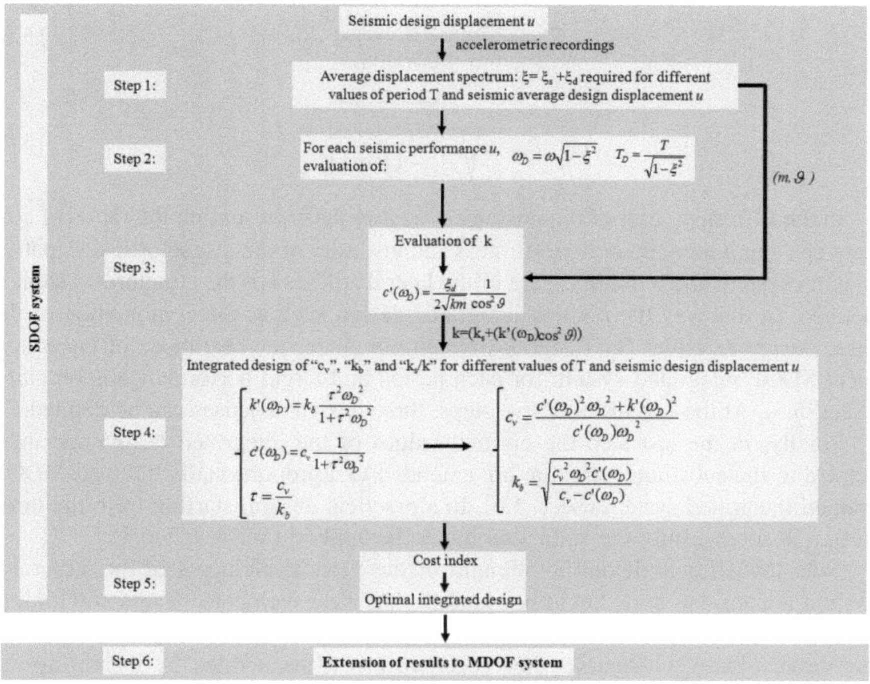

Fig. 5.5 Flow chart of the integrated design methodology

By considering an overall mass m and an inclined angle ϑ of the viscous damper-brace component and using the Eq. (5.5), in the third step, k and $c'(\omega_{\mathrm{D}})$ can be evaluated by deducting the rate representative of the inherent damping of the structural system, taken equal to $\xi_s = 2\ \%$, from the overall viscous damping ratio.

$$c'(\omega_D) = \xi_d \frac{2\sqrt{km}}{\cos^2 \vartheta} \tag{5.5}$$

The fourth step, for each considered vibration period, consists of continuously varying the lateral stiffness of the structural system k_s in the range $[(20/100\ \%)k]$ and, consequently, also the dynamic stiffness of the dissipative bracing system $(k'(\omega_D)\cos^2\vartheta) = (k - k_s)$, in order to obtain for each period and performance the static moduli of the viscous damper-brace component: c_v and k_b. The lateral stiffness of the structural system is considered variable between 20 % and 100 % of the total lateral one and, assigned a value to that variable, its complement to 100 is equal to the dynamic stiffness of the passive control system, connected to the structural system in parallel. The value of 20 % was chosen as a minimum value $k_{s,\min}$ necessary for supporting the gravity loads, as indicated in several scientific studies (Cimellaro et al. 2009; Soong and Cimellaro 2009; Symans et al. 2008). By solving the Eqs. (4.41) and (4.42), the above-mentioned values can be obtained in closed form through the following relations:

$$c_v = \frac{c'(\omega_D)^2\omega_D^2 + k'(\omega_D)^2}{c'(\omega_D)\omega_D^2} \tag{5.6}$$

$$k_b = \sqrt{\frac{c_v^2\omega_D^2 c'(\omega_D)}{c_v - c'(\omega_D)}} \tag{5.7}$$

In the fifth step, a set of dimensionless cost ratios representing the ratios of cost between a unit increase both in stiffness and viscosity of the viscoelastic dissipative bracing system and a unit increase of the lateral stiffness of the structural system is defined. In this way the optimal integrated design such as the combination of the three design variables (k_s, c_v and k_b) which minimizes the cost index of the equivalent SDOF integrated system for each period and target performance is obtained (Sect. 5.5). At the end of these five steps, three design abacuses can be defined.

Finally, in the last step the optimal values of the integrated design variables related to the substitute structure are extended to a proportionally damped MDOF framed integrated system (Sect. 5.6). In a practical design, starting from the three design abacuses, only the sixth step has to be applied (Sects. 6.7–6.8).

Note that, if a limit on the strength of the vertical elements of the structural system is considered, it should be checked that these elements, designed with that optimal value of lateral stiffness, have adequate strength necessary for supporting the gravity loads. Otherwise, the procedure has to be iterated by increasing the lower limit value of the lateral stiffness of the structural system.

If a technological constraint on the overall damping ratio is considered, the methodology can be applied by explicitly considering such a limit. In this case, if a design performance can't be achieved, the period T can be considered between the design variables, too.

The proposed methodology is purely numerical since the problem, due to the non-linearity, cannot be solved in closed form. However, investigating such a wide range of parameters, aims at reaching the qualitative assessment of the role of different variables in the optimization process and the definition of design guidelines and abacuses for practical utility.

In order to apply the proposed integrated design, it should be specified that the frequency, which is considered in the dynamic response of the viscous damper-brace component, is the damped frequency of the response of the single-degree-of-freedom integrated system, since the system filters the seismic signal and presents a dynamic response having energy concentrated on its damped natural frequency. This explains the choice in the previous chapter to use the frequency of the integrated system as a damped frequency of the harmonic displacement function.

It is also highlighted that all the considered design variables, as it can be deduced from the relations (5.6) and (5.7), for a given seismic design displacement and period, are directly proportional to the mass and, therefore, the procedure was carried out in dimensionless terms or with reference to a unit mass.

5.4.1 Accelerometric Recordings Selection

As previously mentioned, the study is carried out on a single-degree-of-freedom integrated system by means of dynamic analyses by selecting from the European database a set of seven historical unscaled accelerometric recordings compatible (CEN 2005) in average with the elastic spectrum (with viscous damping $\xi = 5$ %) relative to the ultimate limit state ULS, life safe, provided by the new Italian seismic code NTC08 (NTC 2008) and considering an ordinary structure on soil type A with a nominal life of 50 years (which corresponds to design for a 475 years return period according to the code) located in Sant'Angelo dei Lombardi (15.1784° longitude, 40.8931° latitude; close to Avellino in Southern Italy).

In particular, the choice of the set of seven acceleration records was conducted by specifying, consistent with the results of the probabilistic seismic hazard analysis (Bazzurro and Cornell 1999), the magnitude (M) and epicentral distance (R) intervals to [6, 7] and [5, 30 km] respectively. In Fig. 5.6, the disaggregation of PGA hazard on rock with a 10 % probability of exceedance in 50 years related to the specified magnitude (M) and distance epicentral (R) intervals is shown.

By using a set of seven acceleration records, according to NTC08, CEN 2005 and Provisions (BSSC, 2004), considering the mean effects on the structure, rather than the maxima, is possible. In fact, a minimum of three accelerograms is required for dynamic linear or non-linear analyses, even though it is usually useful to employ at least seven or more earthquakes. The main advantage of using seven

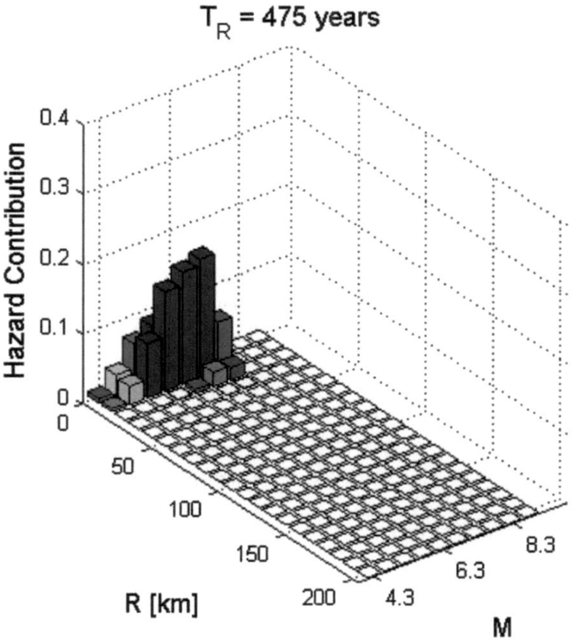

Fig. 5.6 Disaggregation of PGA hazard by considering an ordinary structure located in Sant'Angelo dei Lombardi on rock with a 10 % probability of exceedance in 50 years and specifying the magnitude (M) and distance epicentral (R) intervals to [6, 7] and [5, 30 km] respectively

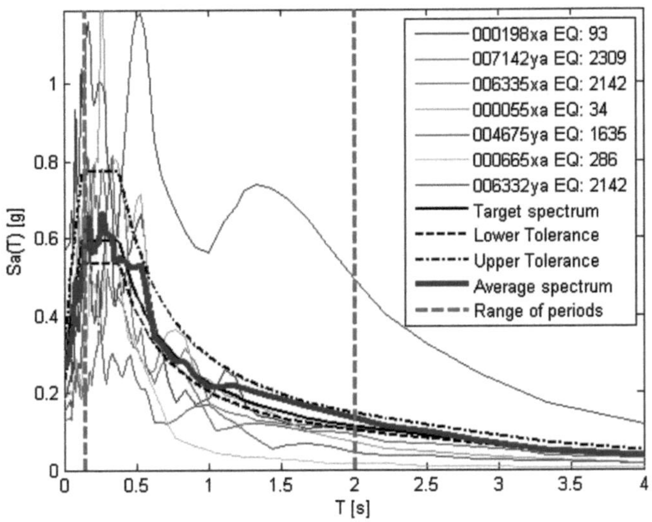

Fig. 5.7 Reference spectrum and spectra of the acceleromtric recordings

or more accelerograms is that the response system can be evaluated by considering the mean effects of the seven responses, instead, if less than seven accelerograms are performed, the maximum values among all the analyses must be considered.

In Fig. 5.7, the elastic spectra (with viscous damping $\xi = 5\ \%$) of the seven selected acceleration records, their average spectrum, the reference spectrum and other two spectra, obtained by assigning a tolerance compatibility of the average,

Table 5.3 Acceleromtric recordings selected

Waveform ID	Station ID	Earthquake Name	Date	M_w	Fault Mechanism
198	ST64	Montenegro	15/04/1979	6.9	Thrust
7142	ST539	Bingol	01/05/2003	6.3	Strike slip
6335	ST2557	South Iceland (after shock)	21/06/2000	6.4	Strike slip
55	ST20	Friuli	06/05/1976	6.5	Thrust
4675	ST2487	South Iceland	17/06/2000	6.5	Strike slip
665	ST238	Umbria Marche	26/09/1997	6	Normal
6332	ST2483	South Iceland (after shock)	21/06/2000	6.4	Strike slip

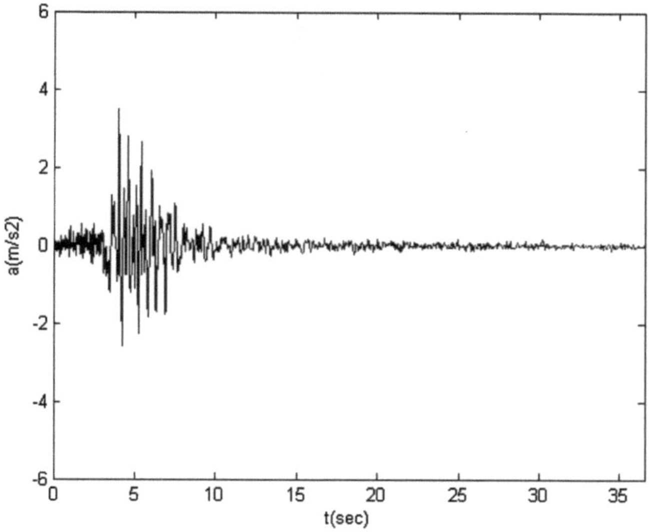

Fig. 5.8 000055xa-Friuli earthquake

respectively, 10 % lower and 30 % upper (CEN 2005; Iervolino et al. 2009), are represented. The spectrum compatibility has been searched in the range of periods 0.15–2 s. Table 5.3 gives a brief description of the selected earthquake events.

The "time-histories" of the seven acceleration records are shown in Figs. 5.8, 5.9, 5.10, 5.11, 5.12, 5.13, 5.14.

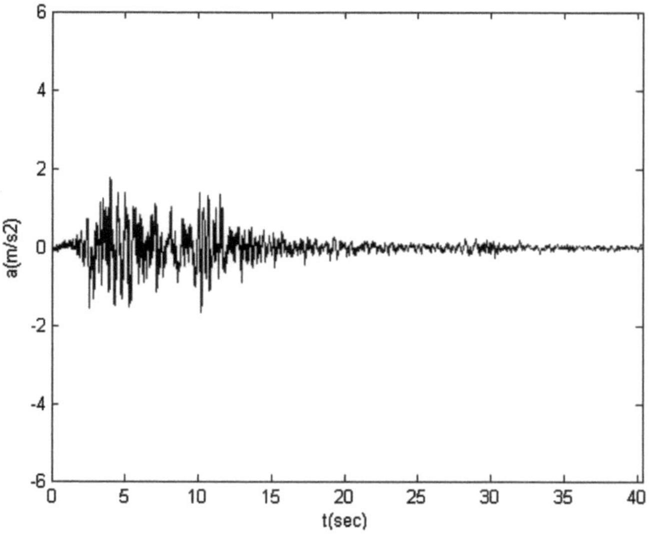

Fig. 5.9 000198xa-Montenegro earthquake

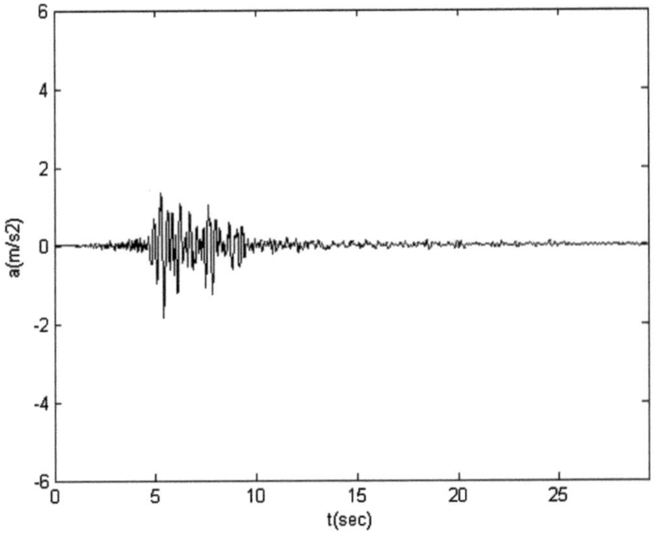

Fig. 5.10 000665xa-Umbria Marche earthquake

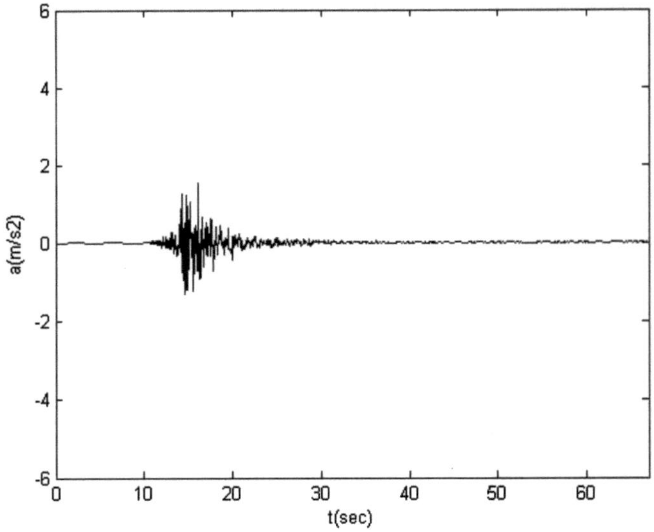

Fig. 5.11 004675ya-South Iceland earthquake

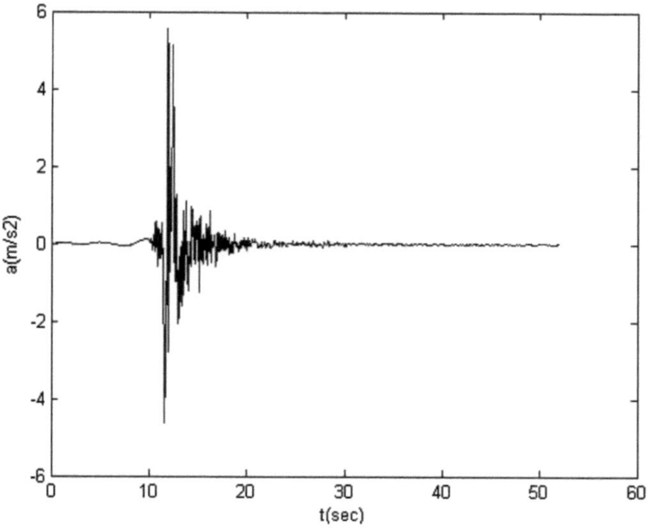

Fig. 5.12 006332ya-South Iceland (after shock) earthquake

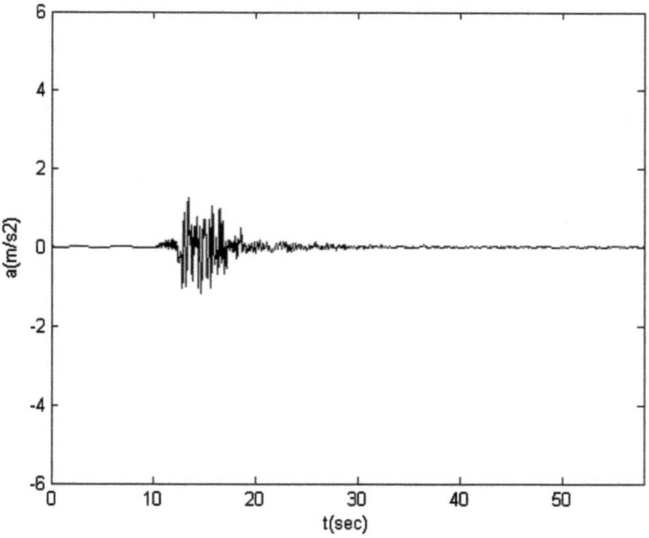

Fig. 5.13 006335xa-South Iceland (after shock) earthquake

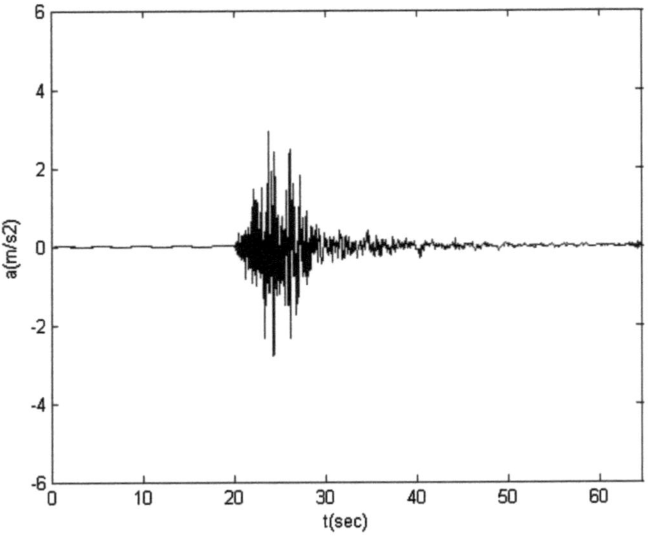

Fig. 5.14 007142ya-Bingol earthquake

5.5 Dimensionless Cost Index

Within the cost-effectiveness in seismic design methods, an optimal combination of design variables should be determined on the basis of the total expected life-cycle cost and the acceptable risk of deaths and potential damage cost from future earthquakes (Park et al. 2004; Ang and Leon 1997). Therefore, an optimal design problem can be formulated as based on a tradeoff between the costs of protection versus potential future losses caused by earthquakes. Life-cycle cost can be defined as the sum of the initial construction cost (García-Pérez 2000; García-Pérez et al. 2007) and the expected damage costs for a given seismic hazard (García-Pérez 2000; Ang and Leon 1997) and the minimization of this index is an optimization problem. In Fig. 5.15, the meaning of the optimal design in economic terms is qualitatively shown.

As for the search of an optimal design, there are different studies in literature which consider explicitly the relationship between the cost of different structural components (Park et al. 2004; Ang and Lee 2001).

In the present work, the following variables have been considered for framed structures equipped with viscoelastic dissipative bracing system in relation to the cost ratios:

- cost C_b for a unit increase of stiffness of the viscoelastic dissipative bracing system divided by the cost C_s for a unit increase of lateral stiffness of the structural system: C_b/C_s;
- cost C_v for a unit increase of the viscosity coefficient of the viscoelastic dissipative bracing system divided by the cost C_s for a unit increase of lateral stiffness of the structural system: C_v/C_s.

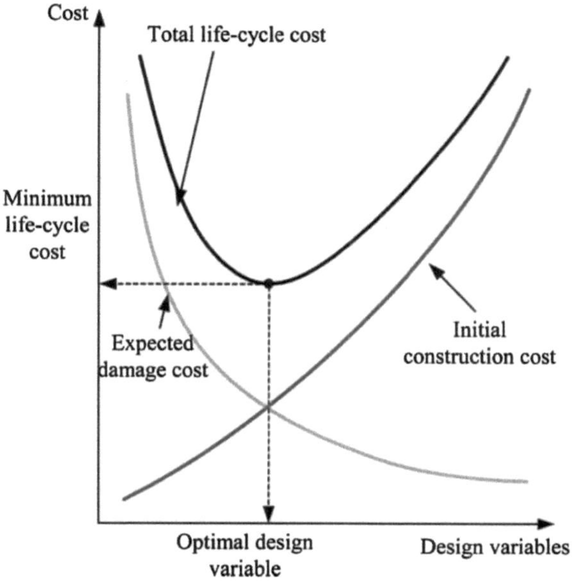

Fig. 5.15 Life-cycle cost minimization concept

Note that the cost of repair, maintenance or damage have not explicitly been expressed as a function of the failure probability of the state limit considered since the achievement of the target performance is a hypothesis of the proposed integrated design.

Different cases of ratios of cost have been chosen by considering the ratio C_b/C_s constant equal to 0.5 and varying only the ratio C_v/C_s, as reported in Table 5.4.

The constant value chosen for the cost ratio C_b/C_s has been established equal to 0.5 since an increase of lateral stiffness can be more conveniently obtained by increasing the cross section of bracing system rather than the bearing vertical elements' ones. Table 5.4 shows that a wide variability range for the ratio C_b/C_s has been analysed since the cost of a viscous linear damper able to provide a design viscosity coefficient depends on several technological parameters, dynamic characteristics of the recordings and structural system (these characteristics

Table 5.4 Pairs of relative ratios of cost

CASE	$\frac{C_b}{C_s}$	$\frac{C_v}{C_s}$
1	0.5	2
2	0.5	3
3	0.5	5
4	0.5	10
5	0.5	25
6	0.5	50
7	0.5	100

identify the maximum displacement, velocity and force which develop between the ends of the device) and many logistical issues including the country where the damped structural system will be built.

By considering the values of the variables corresponding to the average displacement spectrum of the seven accelerometric recordings (CEN 2005), the optimal design values are those which minimize the cost index (5.8), assumed as optimized objective function, of the integrated system. If the Eq. (5.8) is divided by the cost of the structural system, the normalized cost index can be obtained (5.9).

$$C_{tot} = C_s k_s + C_b k_b + C_v c_v \tag{5.8}$$

$$\overline{C}_{tot} = \frac{C_{tot}}{C_s} = k_s + \frac{C_b}{C_s} k_b + \frac{C_v}{C_s} c_v \tag{5.9}$$

The normalized cost of the unbraced structural system and the normalized cost of the braced structure without dampers are, respectively, defined as:

$$\overline{C}_s = \frac{C_s}{C_s} k_s = k_s \tag{5.10}$$

$$\overline{C}_{sb} = \frac{C_s}{C_s} k_s + \frac{C_b}{C_s} k_b \tag{5.11}$$

5.6 Proportionally and Viscoelastically Damped MDOF Framed Integrated Systems

Following the proposed integrated design approach (Fig. 5.5), the optimal values of the integrated design variables related to the substitute structure can be extended to a proportionally damped MDOF framed integrated system, whose dynamic behavior can be characterised by using an equivalent SDOF (Shibata and Sozen 1976; Priestley et al. 1996). With reference to a viscoelastically damped and shear-type framed structural system, the hypotheses, upon which the equivalence between the MDOF framed integrated system and its corresponding substitute structure is based, are:

1. MDOF integrated system characterised by mass and stiffness regularity in elevation;
2. equality between the period and overall damping ratio of the substitute structure and the ones of the first mode of the MDOF integrated system;
3. proportionally damped MDOF integrated system;
4. the economically optimal ratio k_s/k of the equivalent SDOF system must be achieved for each storey of the MDOF integrated system (regular distribution of stiffness);

5. the economically optimal value of the relaxation time τ of the equivalent SDOF system must be achieved for each storey of the MDOF integrated system;
6. the dynamic terms of the stiffness and damping matrices of the viscoelastic dissipative bracing system are evaluated assuming that they only depend on the fundamental frequency of the first mode.

In the above hypothesis the first mode is dominant and there is a straight equivalence between the dynamic behavior of the MDOF system and its equivalent SDOF. The drift profile is thus set by the first mode shape.

In order to extend the proposed procedure to a MDOF system, it is necessary to refer to a design performance u which has to be divided by the first mode participation factor g_1 of the MDOF integrated system (Shibata and Sozen 1976). It follows that the design damping ratio, the economically optimal ratio k_s/k, the economically optimal value of the relaxation time τ and the fundamental damped frequency have to be related to the substitute structure able to achieve the design performance u/g_1.

Finally, it's important to highlight that the choice of the design displacement is generally governed by the limit strains of the critical members. For example, the design displacement for frame buildings will normally be governed by drift limits in the lower storeys of the building. It follows that the methodology can be applied by choosing smaller design displacements in order to respect the drifts limits in accordance to the codes.

On the basis of the above-mentioned hypotheses, the results obtained through the proposed integrated design procedure can be extended to a MDOF framed integrated structure as illustrated in Sects. 6.7 and 6.8. In fact, a proportionally damped MDOF framed integrated system, equipped with "viscous damper-brace component" at each level and designed according to the principles explained in these chapters, can be easily modeled through a substitute structure (Sects. 6.1–6.6) in order to search the optimal values of the design variables. The hypothesis of proportionally damped integrated structural system (i.e., Rayleigh damping) allows to evaluate the dynamic response of the integrated system by considering the dynamic behavior of the viscous damper-brace component and taking into account the presence of the stiffness of the brace of each level or rather the dependence on the product between the damped frequency of the first mode of the (MDOF) integrated system and the relaxation time corresponding to each structural level has been considered to evaluate the dynamic response of each structural level.

On the basis of the above-mentioned hypotheses, the proposed integrated design procedure allows to obtain an economically optimal design of a proportionally damped MDOF framed integrated structure equipped with viscous dampers, in other terms, the least expensive regular system.

5.6.1 Future Developments

The proposed procedure can be extended to other different proportionally and viscoelastically damped elastic earthquake resistant systems, which respect the regularity hypotheses and equipped, in parallel, with a control system which consists of viscoelastc bracing-damper system to dissipate the seismic energy. It follows that the value minimum of the lateral stiffness of the structural system should be modified depending on the type of the lateral loading support system.

The proposed methodology can also be extended to inelastic earthquake resistant systems equipped with a viscoelastic bracing-damper system, which respect the regularity hypotheses and are able to dissipate seismic energy. On the basis of a design hypothesis, the overall viscous damping ratio developed by the integrated system should be devided in three rates:

- inherent damping ξ_s of the structural system;
- supplemental damping ξ_d of the viscoelastic bracing-damper system
- equivalent supplemental damping $\xi_{H,eq}$ of the inelastic system.

The equivalent viscous damping ratio represents the energy dissipated through hystretetic loops by the inelastic earthquake resistant system (Priestley et al. 2007). It follows that other different cost ratios related to the inelastic components should also be defined in order to search the optimal inelastic and elastic design variables.

References

Ang, A.H.-S., Lee, J.-C.: Cost optimal design of R/C structures. Reliabil. Eng. Syst. Saf. **73**, 233–238 (2001)

Ang, A.H.-S., Leon, D.D.: Determination of optimal target reliabilities for design and upgrading of structures. Struct. Saf. **19**(1), 91–103 (1997)

Applied Technology Council (ATC): Seismic evaluation and retrofit of concrete buildings, ATC-40, Redwood City, California (1996)

Applied Technology Council (ATC): NEHRP commentary on the guidelines for the seismic rehabilitation of buildings. Report No. FEMA-273, Prepared for the Building Seismic Safety Council (BSSC) by the ATC, Federal Emergency Management Agency (FEMA), Washington D.C (1997a)

Applied Technology Council (ATC): NEHRP commentary on the guidelines for the seismic rehabilitation of buildings. Report No. FEMA-274, Prepared for the Building Seismic Safety Council (BSSC) by the ATC FEMA, Washington D.C (1997b)

Applied Technology Council (ATC): Prestandard and Commentary for the Seismic Rehabilitation of Buildings, Report No. FEMA-356, Prepared for the Building Seismic Safety Council (BSSC) by the ATC FEMA, Washington D.C (2000)

Bazzurro, P., Cornell, C.A.: Disaggregation of seismic hazard. Bull. Seismol. Soc. Am. **89**, 501–520 (1999)

Building Seismic Safety Council (BSSC): NEHRP recommended provisions for seismic regulations for new buildings and other structures. Report Nos. FEMA-450/1 and FEMA-450/2 (2003 Ed.), prepared by the BSSC for the FEMA, Washington D.C (2004)

CEN–European Committee for Standardization: Eurocode 8 Part 1: General rules, seismic actions and rules for buildings. CEN, Brussels (2005)

Chrysanthakopoulos, C., Bazeos, N., Beskos, D.E.: Approximate formulae for natural periods of plane steel frames. J. Constr. Steel Res. **62**, 592–604 (2006)

Cimellaro, G.P., Soong, T.T., Reinhorn, A.M.: Integrated design of controller linear structural systems. J Struct. Eng. **135**(7), 853–862 (2009)

Crowley, H., Pinho, R.: Revisiting Eurocode 8 formulae for periods of vibration and their employment in linear seismic analysis. Earthquake Eng. Struct. Dynam. **39**(2), 223–235 (2009). doi:10.1002/eqe.949

Ferraioli M., Laezza G., Mandara A.: Impiego di controventi dissipativi per la protezione sismica di strutture in c.a., XX CTA (2005)

Freeman S.A.: Development and use of capacity spectrum method. In: Proceedings of the 6th National Conference on Earthquake Engineering, Seattle (1998)

García-Pérez, J.: Seismic zoning for initial- and total-cost minimization. Earthquake Engineering & Structural Dynamics, **29**, 847−865 (2000)

García-Pérez, J., Zenteno, M., Díaz, O.: Vulnerability functions and the influence of seismic design parameters on initial costs for buildings provided with hysteretic energy-dissipating devices. Earthquake Resistant Engineering Structures, **93**(6), 3−12 (2007). doi: 10.2495/ERES070011

Ghobarah, A.: Performance-based design in earthquake engineering: state of development. Eng. Struct. **23**, 878–884 (2001)

Iervolino, I., Galasso, C., Cosenza, E.: Rexel: computer aided record selection for code-based seismic structural analysis. Bull. Earthquake. Eng. (2009). doi:10.1007/s10518-009-9146

Jalali, A., Milani, A.-S.: Vibration Properties of Steel-Framed Buildings determined from Ambient Vibration Tests. In: 13th World Conference on Earthquake Engineering, August 1–6. Vancouver, Canada (2004)

Kim, J., Choi, H.: Displacement-based design of supplemental dampers for seismic retrofit of a framed structure. J Struct. Eng. **132**(6), 873–883 (2006)

Kim, J.K., Seo, Y.I.: Seismic design of low-rise steel frames with buckling-restrained braces. Eng. Struct. **26**(5), 543–551 (2004)

Kim, J.K., Choi, H.H., Min, K.W.: Performance-based design of added viscous dampers using capacity spectrum method. J. Earthquake Eng. **7**(1), 1–24 (2003)

Lin, Y.Y., Tsai, M.H., Hwang, J.S., Chang, K.C.: Direct displacement-based design for building with passive energy dissipation systems. Eng. Struct. **25**(1), 25–37 (2003)

Miranda, E., Bertero, V.V.: Evaluation of Strength Reduction Factors. Earthquake Spec. Earthquake Eng. Res. Inst. **10**(2), 357–379 (1994)

NTC08: Norme tecniche per le costruzioni, Gazzetta Ufficiale del 04.02.08, DM 14.01.08, Ministero delle Infrastrutture (2008)

Park, K.S., Koh, H.M., Hahm, D.: Integrated optimum design of viscoelastically damped structural systems. Eng. Struct. **26**(5), 581–591 (2004)

Priestley, M.J.N., Seible, F., Calvi, G.M.: Seismic Design and Retrofit of Bridges. John Wiley and Sons, NY (1996)

Priestley, M.J.N., Calvi, M.C., Kowalsky, M.J.: Displacement-Based Seismic Design of Structures. IUSS Press, Pavia (2007)

Rama, Raju K., Ansu, M., Iyer, N.R.: A methodology of design for seismic performance enhancement of buildings using viscous fluid dampers. Struct. Control Health Monit. (2013). doi:10.1002/stc.1568

Ramana, V.G.: Optimum Design of Space Structures with Active and Passive Damping. Engineering with Computers, Springer-Veflag New York Inc. **6**, 177−183 (1990)

Shibata A., Sozen M.A.: Substitute structure method for seismic design in R/C. J Struct. Div. ASCE, **102**(ST1), 1–18 (1976)

Smith, M.J., Grigoriadis, K.M., Skelton, R.E.: Optimal mix of passive and active control in structures. Journal of Guidance, Control, and Dynamics, **15**(4), 912–919, (1992).doi: 10.2514/3.20924

Soong, T.T., Cimellaro, G.P.: Future directions in structural control. Struct. Control Health Monitor. **16**(1), 7–16 (2009)

Structural Engineers Association of California (SEAOC): Recommended Lateral Force Requirements and Commentary. SEAOC, Sacramento (1999)

Sullivan, T.J., Calvi, G.M., Priestley, M.J.N., Kowalski, M.J.: The limitations and performances of different displacement based design methods. J. Earthquake Eng. **7**(1), 201–241 (2003)

Symans, M.D., Charney, F.A., Whittaker, A.S., Constantinou, M.C., Kicher, C.A., Johnson, M.W., McNamara, R.J.: Energy dissipation system for seismic applications: current practise and recent developments. J Struct. Eng. **134**(1), 3–21 (2008)

Tremblay, R.: Fundamental periods of vibration of braced steel frames for seismic design. Earthquake Spec. **21**(3), 833–860 (2005)

Xu B.: Integrated optimization of structure and control systems for interconnected building structures subjected to earthquake, J Vibr. Control (2013) doi:10.1177/1077546312469427

Xu, B., Zhang, Y.L., Yao, S.X., Jiang, J.S.: Integrated design of control/structural systems based on genetic algorithm. Appl. Mech. Mater. **204**, 4855–4867 (2012)

Zhang, R.H., Soong, T.T.: Seismic design of viscoelastic dampers for structural applications. J. Struct. Eng. **118**(5), 1375–92 (1992)

Chapter 6
Applying the Optimal Integrated Design Methodology

Abstract This chapter deals with the proposed parametric analysis of the integrated design methodology by considering the set of acceleration records selected in the previous chapter. With reference to an equivalent SDOF integrated system, starting from the evaluation of the average displacement spectrum, combinations of the design variables, which are the lateral stiffness of the structural system, the static stiffness and the static damping coefficient of the viscoelastic dissipative bracing system, are evaluated for different values of the period of the system and of the seismic design displacement considered. Subsequently, the minimum of the cost index is searched for each group of cost ratios in order to find the optimal values of the design variables. This is followed by an economic comparison between the optimal integrated structural/control system and the optimal elastic braced structural system or the optimal elastic unbraced structural system without dampers for each target performance. In the last part, a validation of the proposed procedure is performed by verifying that an optimal single-degree-of-freedom integrated system achieves the expected seismic design displacement. Finally, the extension of this methodology to a proportionally damped multi-degrees-of-freedom framed integrated system is developed on the basis of specific hypotheses to demonstrate the effectiveness of the proposed integrated design methodology.

6.1 Structural Model Employed in Parametric Analyses

In the dynamic analyses carried out to develop the parametric study described in the previous chapter, the equivalence between single-degree-of-freedom (SDOF) integrated systems and the corresponding multi-degrees-of-freedom (MDOF) integrated framed systems was exploited (Shibata and Sozen 1976; Priestley et al.

P. Castaldo, *Integrated Seismic Design of Structure and Control Systems*,
Springer Tracts in Mechanical Engineering, DOI: 10.1007/978-3-319-02615-2_6,

1996). This equivalence does not involve strong approximations in the case of structural systems characterized by a regularity in elevation so that the first mode can be regarded as representative of the dynamic response of the system. Experimental evidence, as already mentioned, (Hwang and Huang 2003; Hwang et al. 2008; Hwang 2005) has shown that if the damping ratio of a structure is increased, the higher mode responses of the structure will be suppressed. It follows that in the case of supplemental damping in regular MDOF structural systems, the hypothesis to refer only to first mode, in the practical applications, can be considered an approximation even more acceptable especially with reference to the new constructions, whose seismic design is based on criteria of structural regularity. In relation to systems irregular in elevation, the combined effects of higher modes can hardly be overlooked because they may be important in order to evaluate the velocities of the plane and, consequently, for the design of the rate-dependent devices. The hypotheses to consider the fundamental mode alone is present in various codes (ATC 1996, 1997a, b; NTC 2008; CEN 2005) in which it is prescribed that the collapse mechanism for the construction is a single-degree-of-freedom mechanism so that the distribution of displacements over the entire height of the building can be reasonably estimated by using both the first mode shape or another linear profile, as an inverted triangle.

Therefore, assuming that the dynamic behavior of a n-degrees-of-freedom framed system regular in elevation, equipped with viscous linear dampers (Fig. 6.1) is governed by the first mode, a single-degree-of-freedom system equipped with the viscous damper-brace component has been employed (Fig. 6.2).

With reference to the single-degree-of-freedom integrated system, therefore, the results of numerical analyses with the aim of investigating the response of the

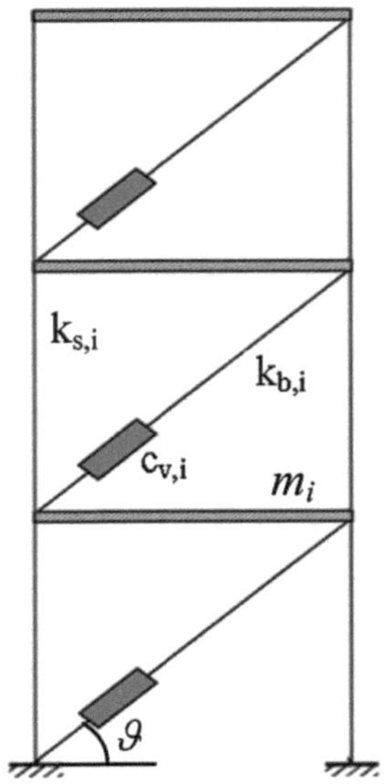

Fig. 6.1 MDOF system equipped with linear viscous dampers

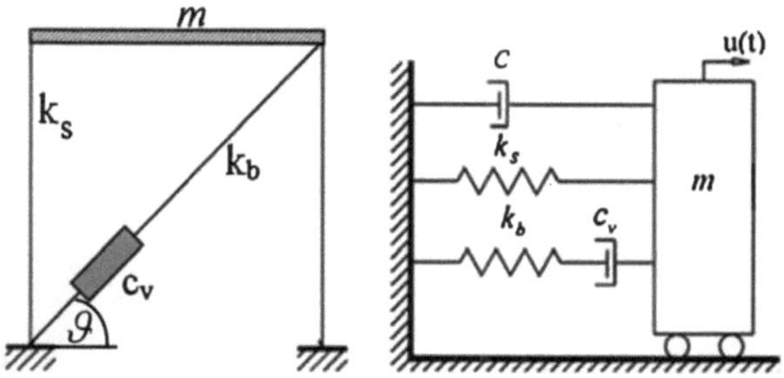

Fig. 6.2 Analytical model of the equivalent SDOF system with viscous damper

elastic structural system equipped with viscoelastic dissipative bracing system, based on the equations presented in Chap. 4 and represented in the state space form (Chap. 3), are described. The numerical algorithm presented in the previous chapter has been implemented in MATLAB by taking into account the set of the historical unscaled accelerometric recordings selected and described in Sect. 5.4.1.

6.2 Evaluation of Viscous Damping Ratio for Each Average Design Displacement

With reference to each of the seven accelerometric recordings, described in Sect. 5.4.1, the damping ratio demanded by each one of the seismic events considered, for different values of the period T and the sesmic design displacements *u*, was evaluated in order to obtain the displacement demand spectra of damping ratio for each average design displacement.

These spectra have been discussed in several literature studies, as illustrated in Miranda and Bertero (1994).

In the analyses, in a first phase, it was chosen to investigate all the variabilities of the supplemental damping ratio with regard to under-damped structures.

In Fig. 6.3a, the viscous damping ratio ξ, of which the rate variable between 2 % and 100 % is the supplemental damping ratio ξ_d, demanded by one of the seismic events considered, for different values of the period T of the integrated structural system and the expected target performance u is illustrated. The contour lines representing the value of the viscous damping ratio ξ, demanded by each earthquake considered, for different values of the period T of the integrated structural system and seismic design displacement u are shown in Figs. 6.3b, 6.4, 6.5, 6.6, 6.7, 6.8 and 6.9.

On the basis of the seven displacement spectra corresponding to the accelerometric recordings, it has been possible to evaluate the average displacement spectrum represented in Fig. 6.10. In Fig. 6.11, the corresponding contour lines are plotted.

The area in which no control is required is the portion of the spectrum with $\xi = \xi_s = 2\,\%$ according to Eq. (5.3). The dashed area represents the portion of the spectrum in which the oscillator is overdamped or critically damped. This area has

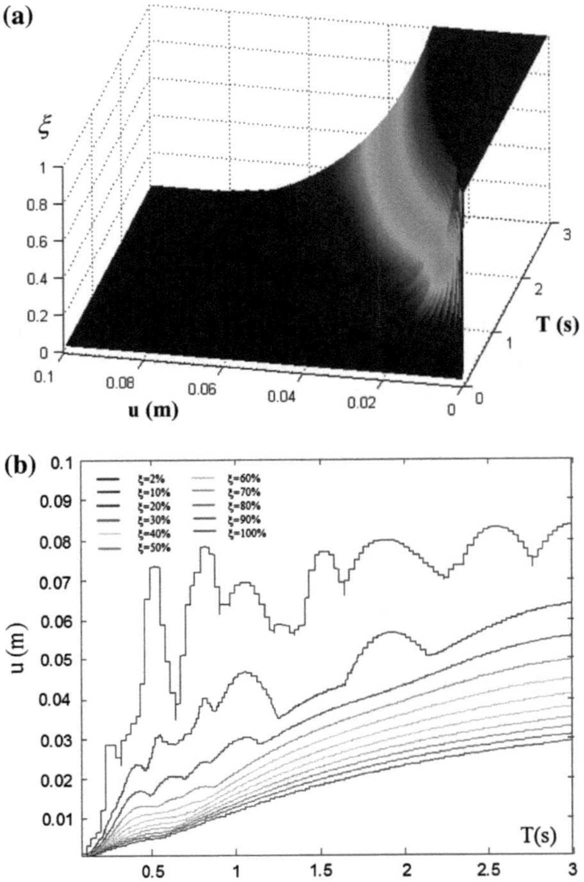

Fig. 6.3 Damping ratio demanded by the 000055xa accelerometric recording (**a**) and contour lines (**b**)

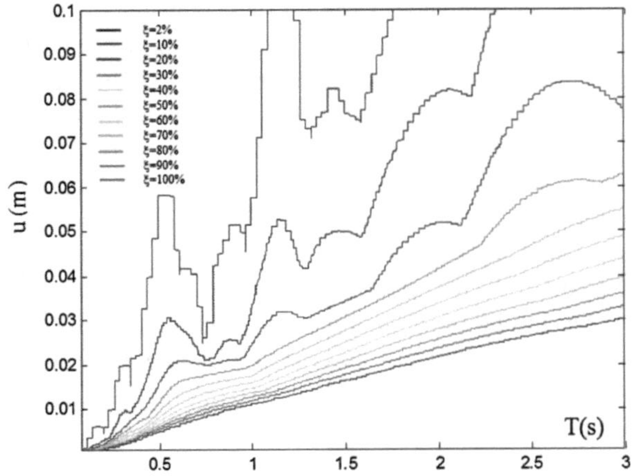

Fig. 6.4 Contour lines of the damping ratio demanded by the 000198xa accelerometric recording

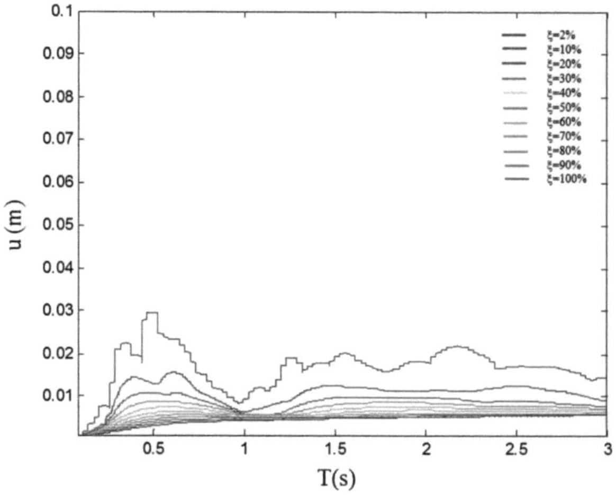

Fig. 6.5 Contour lines of the damping ratio demanded by the 000665xa accelerometric recording

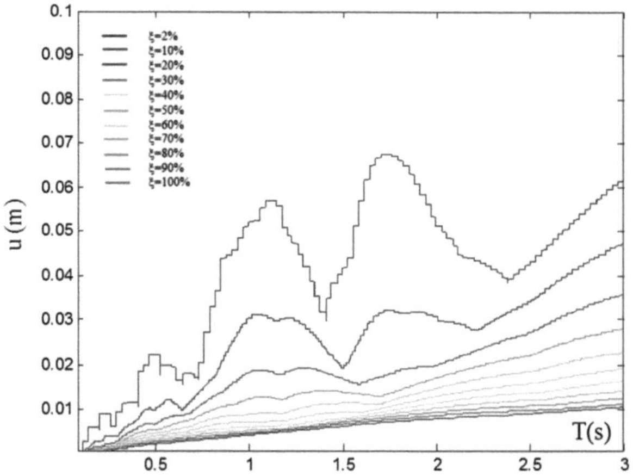

Fig. 6.6 Contour lines of the damping ratio demanded by the 004675ya accelerometric recording

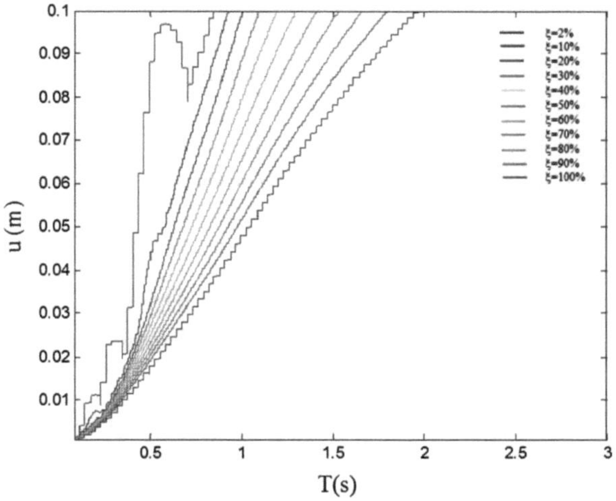

Fig. 6.7 Contour lines of the damping ratio demanded by the 006332ya accelerometric recording

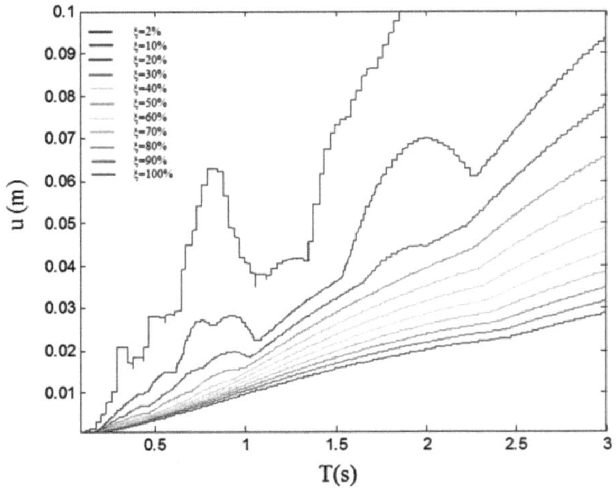

Fig. 6.8 Contour lines of the damping ratio demanded by the 006335xa accelerometric recording

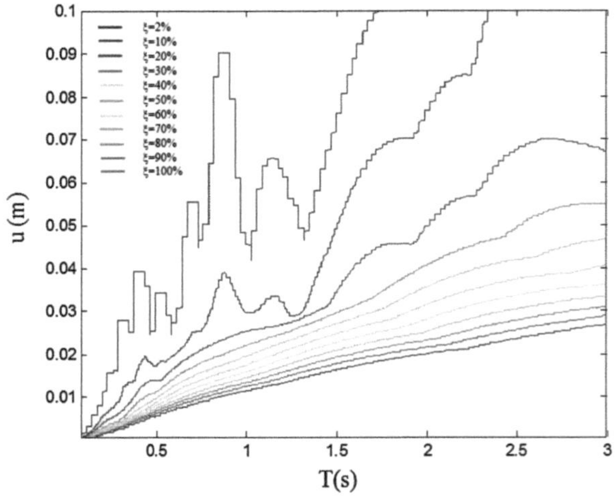

Fig. 6.9 Contour lines of the damping ratio demanded by the 007142ya accelerometric recording

not been taken into account in the analyses. Only (under)damped systems have been considered.

For application, moreover, in a second phase, analyses have also been carried out by imposing a maximum limit on the overall viscous damping ratio ξ equal to 40 %.

6.3 Applying the Integrated Design Methodology by Evaluating the Viscoelastic Design Variables

The evaluation of the average displacement spectrum, represented in Fig. 6.11, presents the first step of the integrated design procedure and defines the design damping ratio for each period and average design performance. The design overall

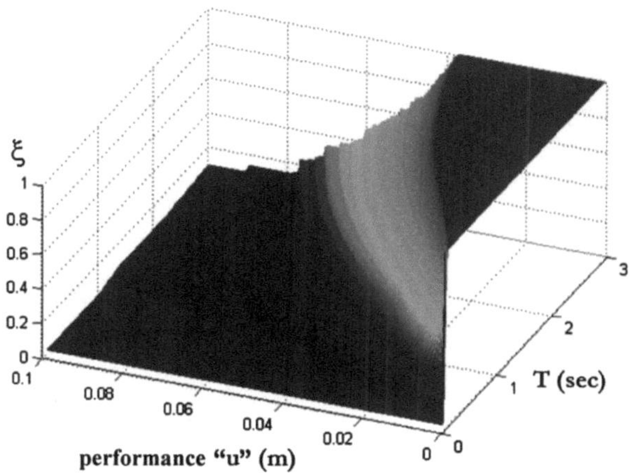

Fig. 6.10 Average displacement spectrum

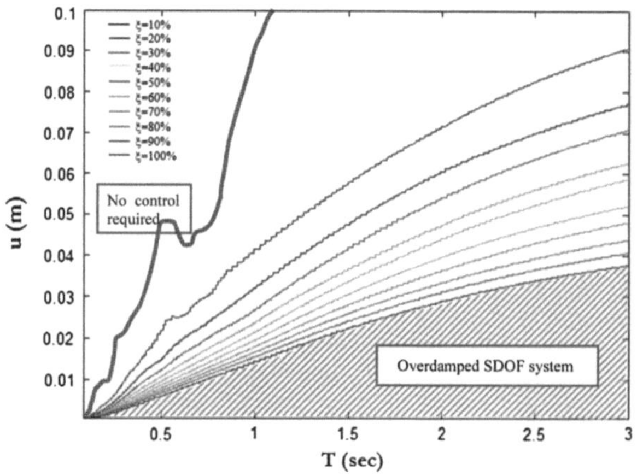

Fig. 6.11 Average displacement spectrum: contour lines

viscous damping ratio is composted of the rate representative of the inherent damping of the structural system, assumed equal to $\xi_s = 2\,\%$, and the one representative of the supplemental damping of the viscoelastic dissipative bracing system, $\xi_d = \xi - \xi_s$, according to Eq. (5.3). By assuming a unit mass m and an inclined angle ϑ of the viscous damper-brace component, which will be always supposed equal to 45°, it has been, therefore, possible to evaluate, for each period T (i.e., undamped natural frequency ω) and design displacement u, the total lateral stiffness of the integrated structural system k and the dynamic damping coefficient of the dissipative bracing system $c'(\omega_D)$ according to Eqs. (5.4) and (5.5).

Note that the seismic design displacement ranges from 1 mm to 10 cm with a very dense pitch equal to one tenth of a millimeter.

Then, continuously varying, with percentage unit changes, the lateral stiffness of the structural system k_s and the dynamic stiffeness of the viscoelastic bracing system connected in parallel $k'(\omega_D)$ for each period and target performance, as

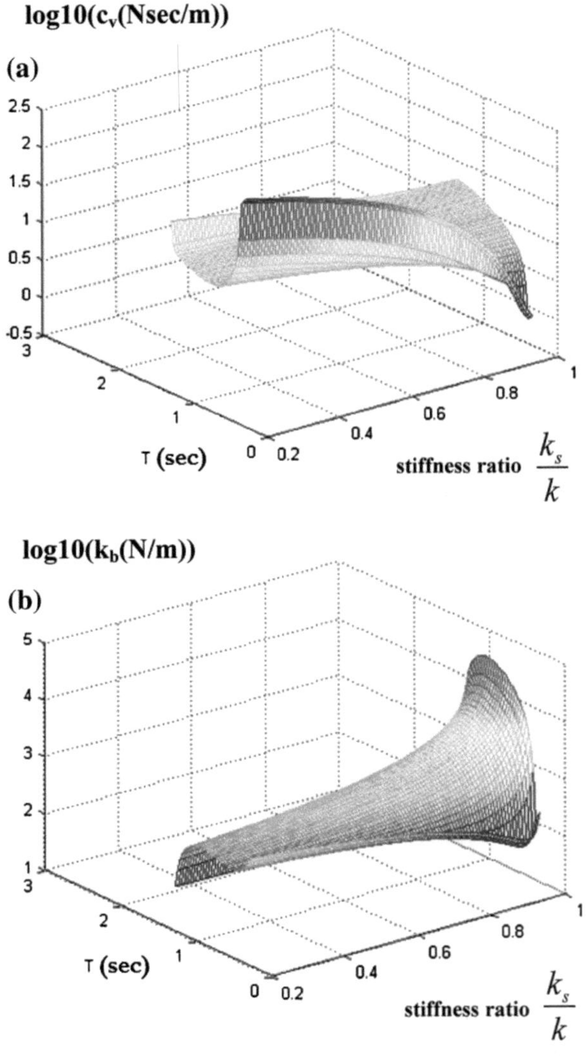

Fig. 6.12 For a seismic design performance $u = 2$ cm, $\vartheta = \pi/4$: **a** c_v; **b** k_b

described in Sect. 5.4, the static moduli of the viscous damper-brace component c_v and k_b have been valuated by using the Eqs. (5.6) and (5.7).

In this section the steps 2–4 of the proposed integrated design methodology (Sect. 5.4, Fig. 5.5) have been developed.

With reference to some of the seismic design displacements analysed, in particular the displacement values: $u = 2, 3, 4, 5, 6, 7$ cm, respectively, in Figs. 6.12, 6.13, 6.14, 6.15, 6.16, and 6.17 the values, expressed in base-10 logarithmic scale, of the two viscoelastic design variables, static stiffness k_b and static damping coefficient c_v of the dissipative bracing system, are represented for different values of the period T and k_s/k, which guarantee in average the achievement of the target performance.

Fig. 6.13 For a seismic design performance u = 3 cm, $\vartheta = \pi/4$: **a** c_v; **b** k_b

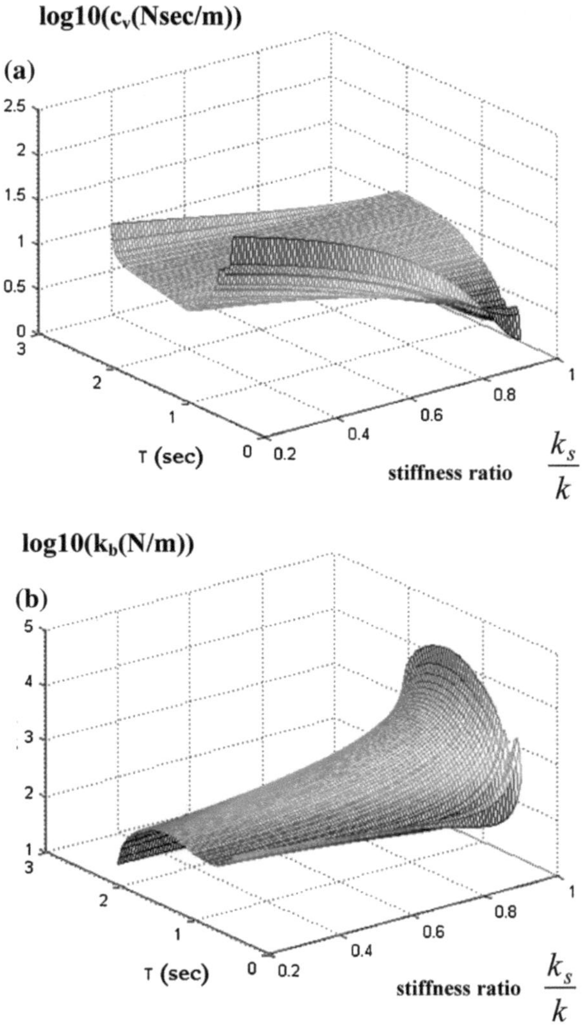

From Fig. 6.12b, 6.13, 6.14, 6.15, 6.16 and 6.17b, it must be highlighted that, on the basis of the reduction coefficients illustrated in Fig. 4.12, when the ratio of the lateral stiffness of the structural system divided by the total one tends to 1, or rather when the dynamic (apparent) stiffness $k'(\omega_D)$ of the viscoelastic dissipative bracing system tends to zero the static stiffness k_b tends to increase significantly, as discussed in Sect. 4.3.

It is also worth noting that the inclination angle, assumed equal to 45°, represents an indicative value that can be highly variable in function of the geometrical characteristics of the structure or geometric configurations of the energy dissipation assemblies for the purpose of amplification (Sigaher and Constantinou 2003) (Sect. 6.3.1).

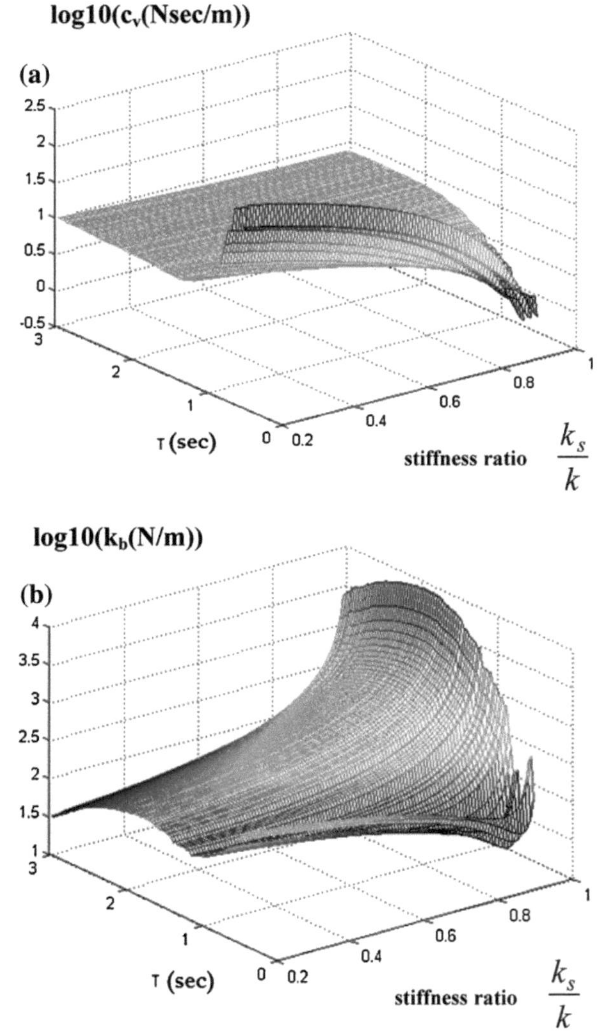

Fig. 6.14 For a seismic design performance $u = 4$ cm, $\vartheta = \pi/4$: **a** c_v; **b** k_b

6.3.1 The Device Configurations

The device configurations either in-line with diagonal bracing or as horizontal elements atop chevron bracing (Sigaher and Constantinou 2003) are the most employed. The popularity of these configurations is based on the engineers' familiarity with such bracing systems and the fact that all experimental research studies have utilized mostly these two configurations for energy dissipation systems. Another device configuration recently developed is the toggle-brace configuration. As the name implies, this configuration is based on the toggle mechanism, which magnifies the damper displacement for a given interstory drift. This amplification results in a reduction in the required damping force, and

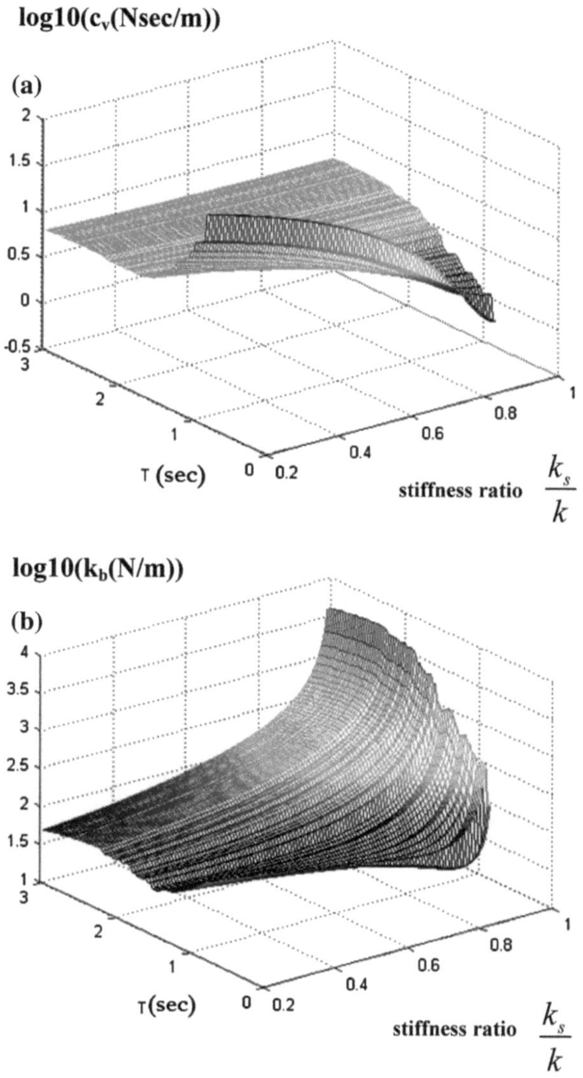

Fig. 6.15 For a seismic design performance $u = 5$ cm, $\vartheta = \pi/4$: **a** c_v; **b** k_b

reduction in the damper volume, which may lead to reduction of damper costs. With the intent that the energy dissipation assemblies do not occupy entire bays in frames and often violate architectural requirements such as open space and unobstructed view, the scissor-jack-damper system was developed as a variant of the toggle-brace-damper system. The system combines the displacement magnification feature with small size, which is achieved through compactness and near-vertical installation. The displacement of the energy dissipation devices is either less than (case of diagonal brace) or equal to (case of chevron brace) the drift of the story at which the devices are installed. If u and u_d denote the interstory drift and the damper relative displacement, respectively, it can be written:

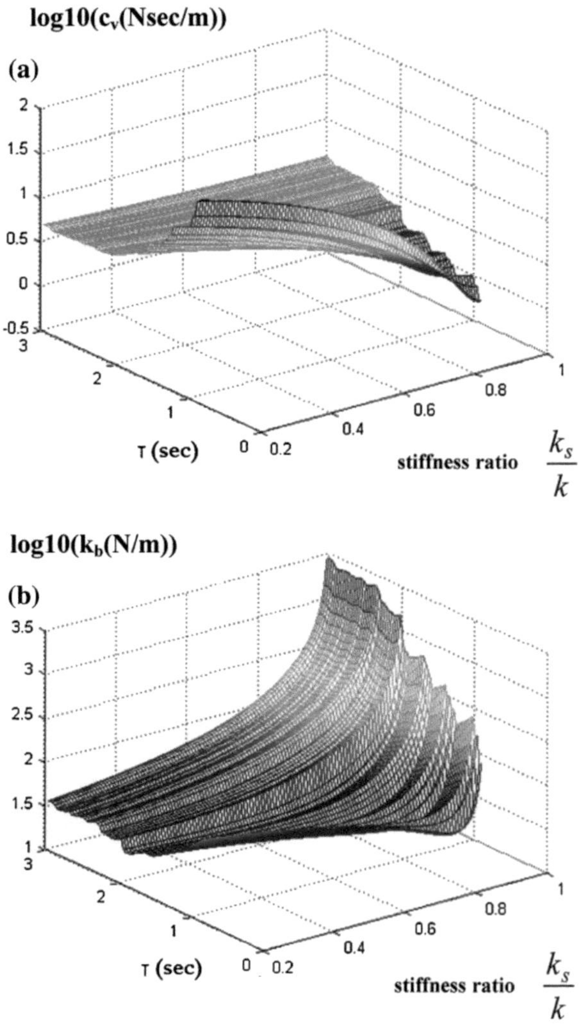

Fig. 6.16 For a seismic design performance $u = 6$ cm, $\vartheta = \pi/4$: **a** c_v; **b** k_b

$$u_d = f \cdot u \tag{6.1}$$

where f is the magnification factor. For the chevron brace configuration, $f = 1$; for the diagonal configuration $f = \cos\vartheta$, where ϑ is angle of inclination of the damper with respect to the horizontal axis. The force F_d along the damper axis is similarly related to F, the horizontal component of the damper force exerted on the frame, through:

$$F = f \cdot F_d \tag{6.2}$$

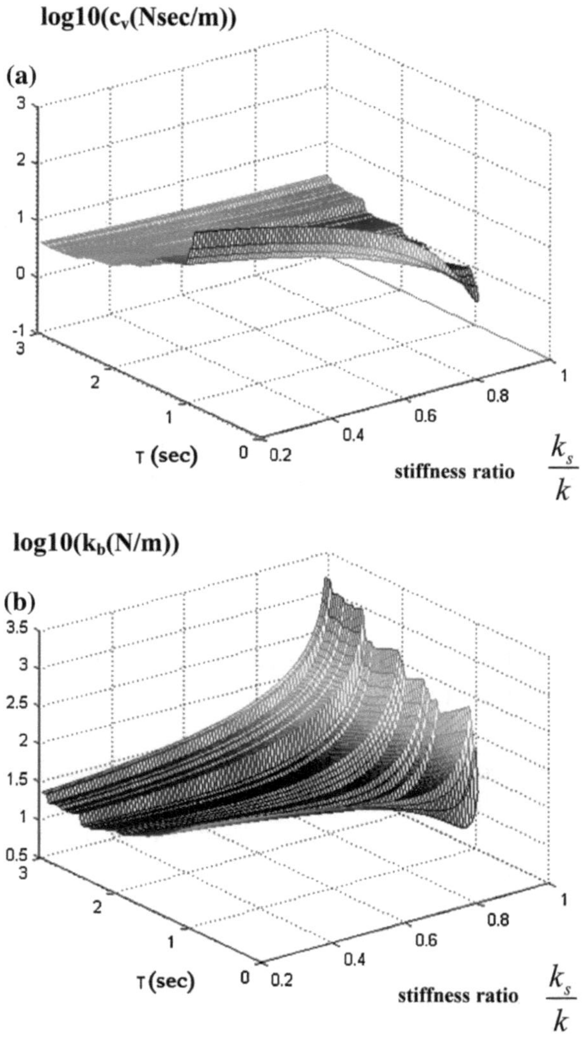

Fig. 6.17 For a seismic design performance u = 7 cm, $\vartheta = \pi/4$: **a** c_v; **b** k_b

Consider a SDOF system having mass *m*, a fundamental vibration period *T* and that is equipped with a linear fluid viscous damper, the resistance force, evaluated through Eq. (2.28), is equal to:

$$F_d = c\dot{u}_d \tag{6.3}$$

where *c* is the viscosity coefficient, and $\dot{u}_d$ relative velocity between the ends of the damper along its axis. The damping force F, exerted on the frame by the damper assembly is given by:

$$F = f \cdot F_d = f \cdot c \cdot \dot{u}_d = f^2 \cdot c \cdot \dot{u} \tag{6.4}$$

in which $\dot{u}$ is interstory velocity. It follows that the supplemental damping ratio of a single-story frame with a linear fluid viscous device can be written as:

$$\xi_d = \frac{cf^2}{2\omega m} \tag{6.5}$$

It is essential to realize the effect of the magnification factor on the damping ratio, in fact, the damping ratio varies proportionally with the square of the magnification factor.

In contrast to the familiar diagonal and chevron brace configurations, the scissor-jack configuration can achieve magnification factors substantially greater than unity. The presence of the magnifying mechanism in the scissor-jack system extends the utility of fluid viscous devices to cases of small interstory drifts and velocities, which are typical of stiff structural systems under seismic excitation and structures subjected to wind load. Assuming a supplemental damping ratio equal to 0.05, developed by a linear fluid viscous damper in chevron configuration, Fig. 6.18 shows the values of both the supplemental damping ratio and magnification factor f for the different configurations.

6.4 The Optimal Design: Evaluation of the Optimal Design Variables

In this section the optimal design of the integrated structural system is evaluated in economic terms regarding the seven cases of relative cost ratios defined on the design variables considered, shown in Table 5.4, by evaluating the minimum of the normalized cost index (Eq. (5.9), in Sect. 5.5), calculated by comparing all the integrated design solutions able to achieve the seismic design displacement. This is the fifth step of the proposed integrated design methodology (Sect. 5.4, Fig. 5.5).

In fact, for each pair of cost ratios, the optimal value of the stiffness of the structural system is chosen corresponding to the optimal values of the static stiffness and static viscosity coefficient of the viscoelastic dissipative bracing system so that these three optimal values of the design variables minimize the cost index considered. The above-mentioned choice is made among all the combinations obtained by continuously varying the stiffness of the structural system and that of the dissipative bracing system, as discussed in Sect. 5.3. The methodology developed, as mentioned in Sect. 5.4, is purely numerical since the problem, due to the non-linearity, cannot be solved in closed form.

With reference to the case 1, $C_b/C_s = 0.5$ and $C_v/C_s = 2$, by considering $\vartheta = \pi/4$ and a unit mass, in Fig. 6.19 the economically optimal value of the ratio of the lateral stiffness of the structural system divided by the total lateral stiffness of the integrated system k_s/k is shown for different values of the period T (i.e., undamped natural frequency ω), and seismic design performance u.

Diagonal	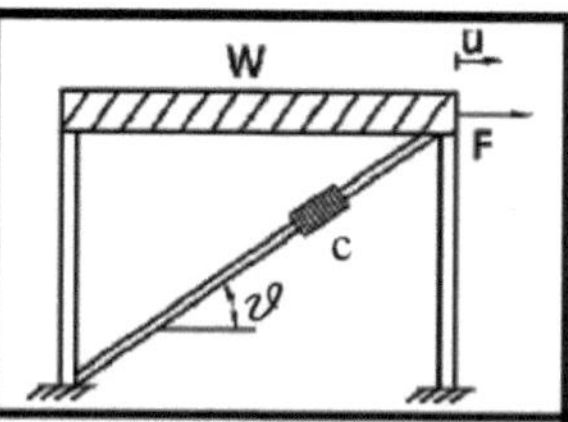	$f = \cos \vartheta$	$\vartheta = 37°$ $f = 0.80$ $\xi_d = 0.03$
Chevron	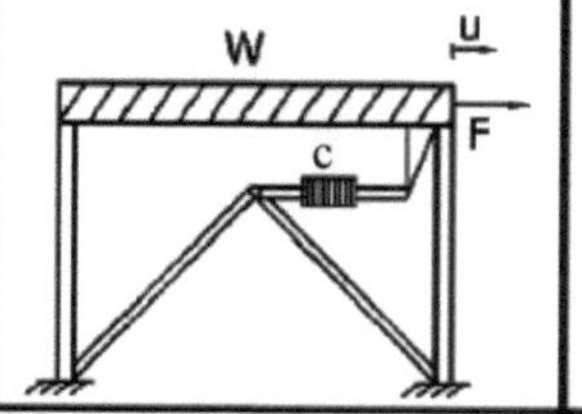	$f = 1$	$f = 1.00$ $\xi_d = 0.05$
Scissor-Jack	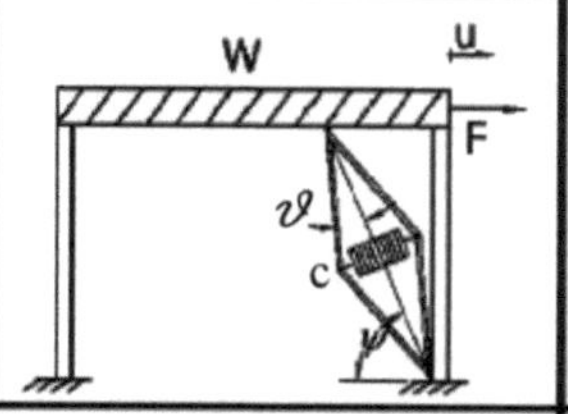	$f = \dfrac{\cos \psi}{\tan \vartheta}$	$\vartheta = 9°$ $\psi = 70°$ $f = 2.16$ $\xi_d = 0.23$
Upper Toggle	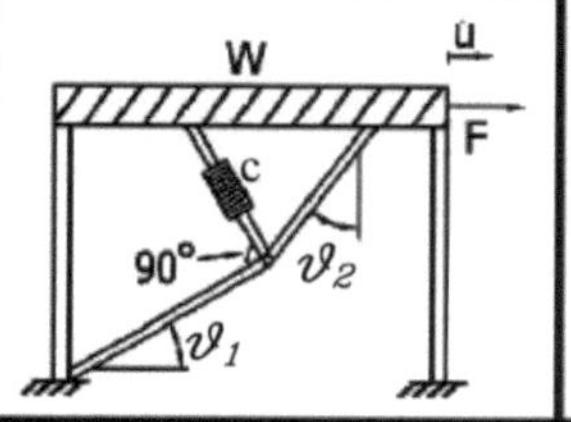	$f = \dfrac{\sin \vartheta_2}{\cos(\vartheta_1 + \vartheta_2)} + \sin \vartheta_1$	$\vartheta_1 = 31.9°$ $\vartheta_2 = 43.2°$ $f = 3.191$ $\xi_d = 0.509$
Reverse Toggle	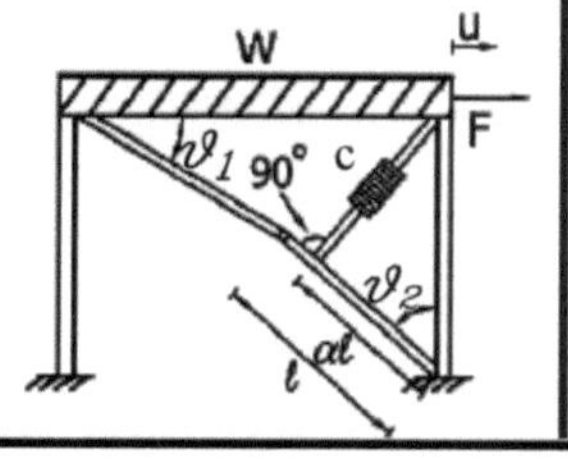	$f = \dfrac{\alpha \cos \vartheta_1}{\cos(\vartheta_1 + \vartheta_2)} - \cos \vartheta_2$	$\vartheta_1 = 30°$ $\vartheta_2 = 49°$ $\alpha = 0.7$ $f = 2.521$ $\xi_d = 0.318$

Fig. 6.18 Illustration of diagonal, chevron brace, scissor-jack-damper, and toggle-brace-damper configurations, magnification factors, and damping ratios of a single-story structure with linear fluid viscous devices (modified from Sigaher and Constantinou 2003)

Figures 6.20 and 6.21 represent, respectively, the optimal values of the static stiffness k_b and the static damping coefficient c_v of the viscoelastic dissipative bracing system corresponding to the optimal values of the ratio of the lateral stiffness of the structural system divided by the total lateral stiffness of the integrated system k_s/k, for different values of the period T (i.e., undamped natural frequency ω), and seismic design performance u.

Fig. 6.19 Optimal ratio k_s/k

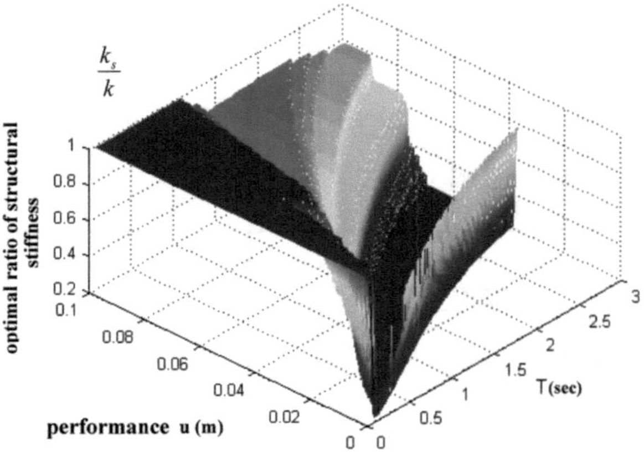

Fig. 6.20 Optimal value of k_b related to optimal ratio k_s/k

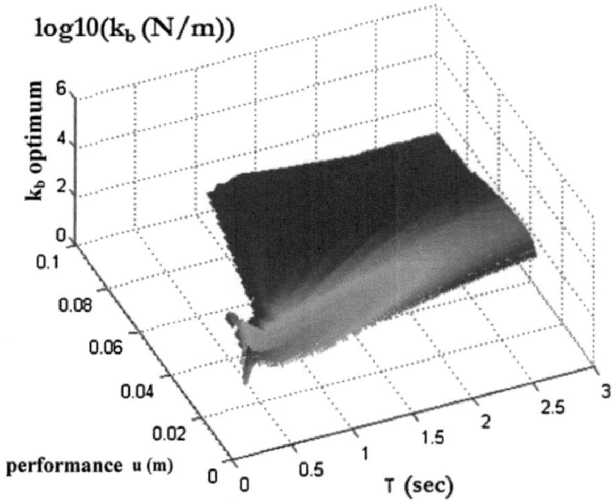

Fig. 6.21 Optimal value of c_v related to optimal ratio k_s/k

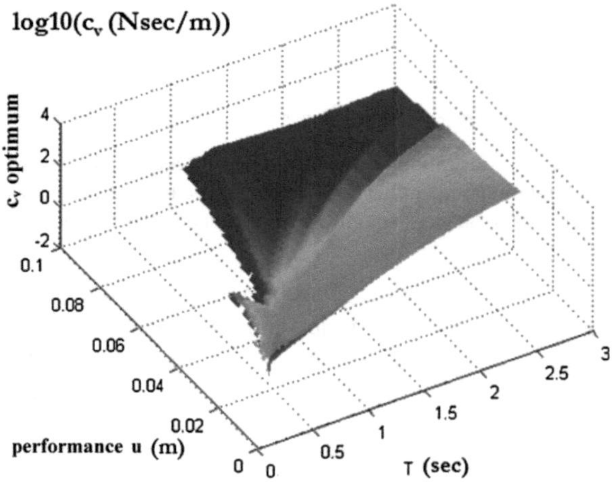

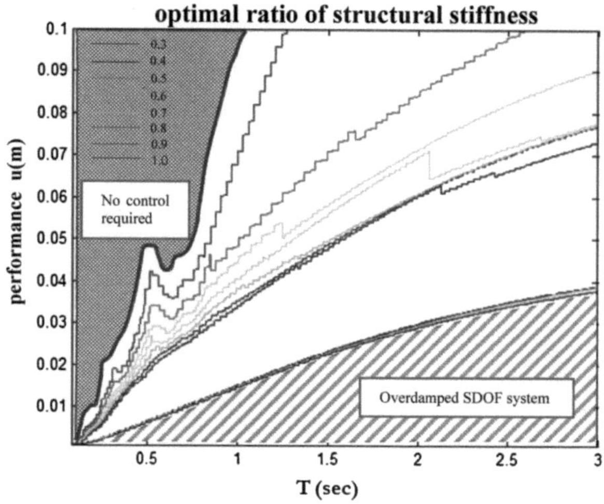

Fig. 6.22 Contour lines of optimal ratio k_s/k

In Figs. 6.22, 6.23 and 6.24, the corresponding contour lines are represented respectively.

Similarly, the results relating to case 2, $C_b/C_s = 0.5$ and $C_v/C_s = 3$, are shown in Figs. 6.25, 6.26, 6.27, 6.28, 6.29 and 6.30.

Similarly, the results relating to case 3, $C_b/C_s = 0.5$ and $C_v/C_s = 5$, are shown in Figs. 6.31, 6.32, 6.33, 6.34, 6.35 and 6.36.

Similarly, the results relating to case 4, $C_b/C_s = 0.5$ and $C_v/C_s = 10$, are shown in Figs. 6.37, 6.38, 6.39, 6.40, 6.41 and 6.42.

Similarly, the results relating to case 5, $C_b/C_s = 0.5$ and $C_v/C_s = 25$, are shown in Figs. 6.43, 6.44, 6.45, 6.46, 6.47 and 6.48.

Similarly, the results relating to case 6, $C_b/C_s = 0.5$ and $C_v/C_s = 50$, are shown in Figs. 6.49, 6.50, 6.51, 6.52, 6.53 and 6.54.

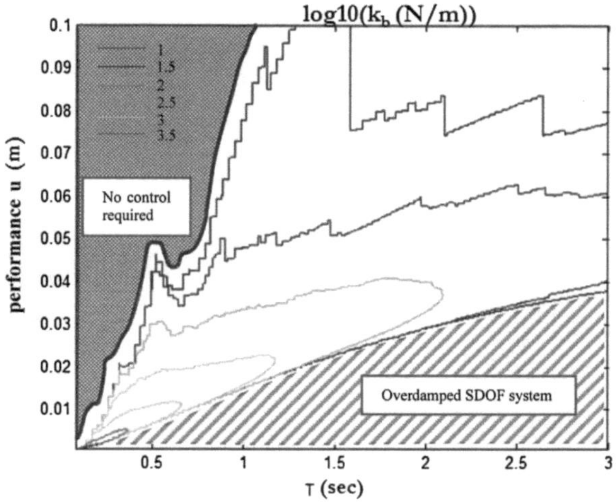

Fig. 6.23 Contour lines of optimal value of k_b related to optimal ratio k_s/k

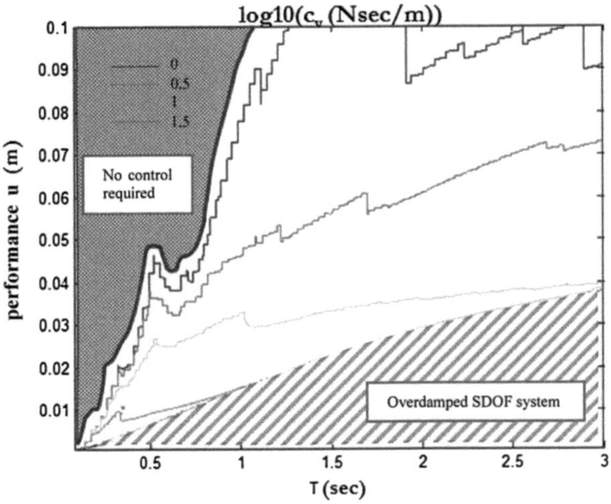

Fig. 6.24 Contour lines of optimal value of c_v related to optimal ratio k_s/k

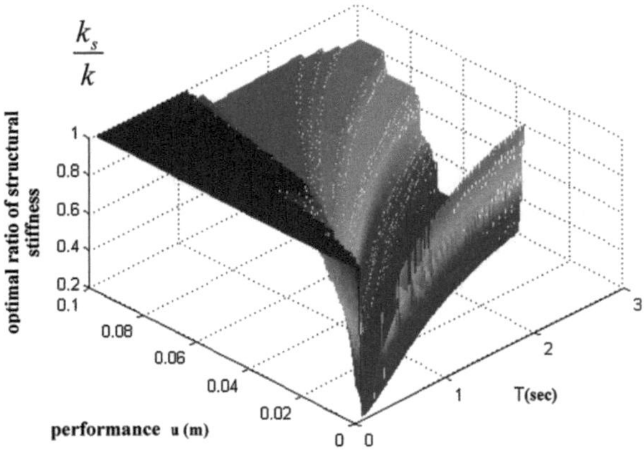

Fig. 6.25 Optimal ratio k_s/k

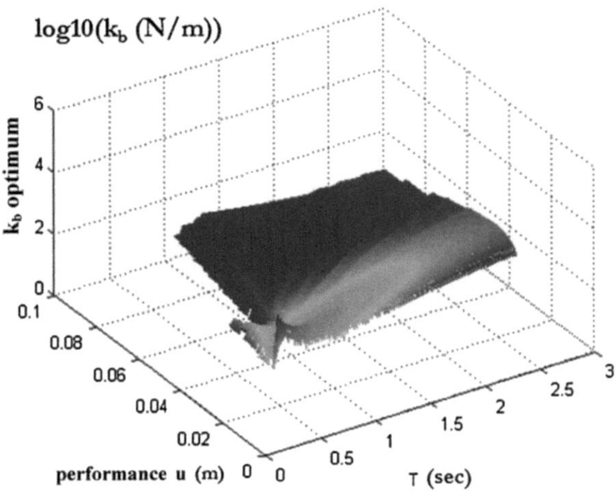

Fig. 6.26 Optimal value of k_b related to optimal ratio k_s/k

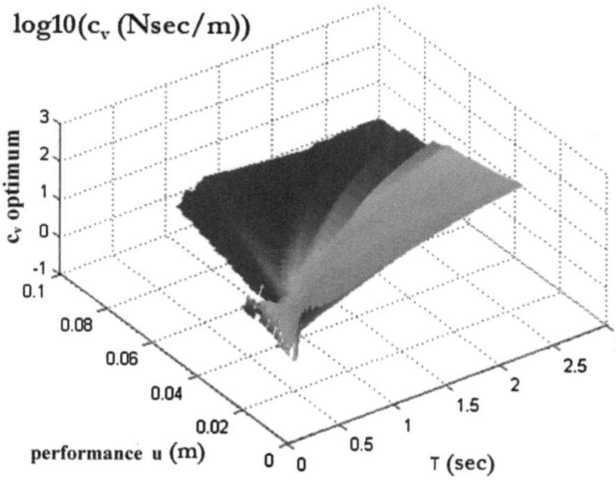

Fig. 6.27 Optimal value of c_v related to optimal ratio k_s/k

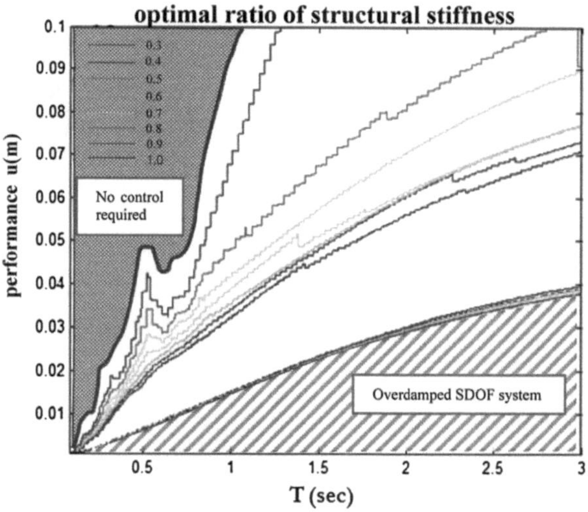

Fig. 6.28 Contour lines of optimal ratio k_s/k

Similarly, the results relating to case 7, $C_b/C_s = 0.5$ and $C_v/C_s = 100$, are shown in Figs. 6.55, 6.56, 6.57, 6.58, 6.59 and 6.60.

For each of the seven cases analysed, three design abacuses and their respective contour lines have been defined in order to design an optimal integrated system located in an area characterized by the seismic intensity corresponding to the reference displacement average spectrum considered. In fact, from these abacuses, it is possible to obtain the economically optimal values of the lateral stiffness of the structural system, the static stiffness and static viscosity coefficient of the viscoelastic dissipative bracing system, for different values of the elastic natural period T and the target performance u.

From Fig. 6.19, 6.25, 6.31, 6.37, 6.43, 6.49 and 6.55, it can be highlighted that the dashed area relative to the very small displacements is the area in which the

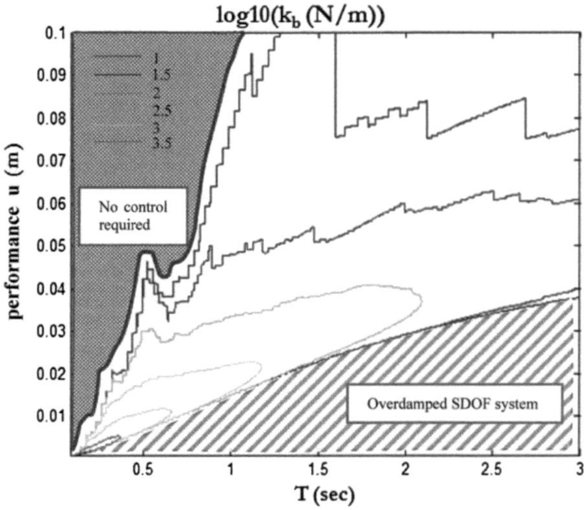

Fig. 6.29 Contour lines of optimal value of k_b related to optimal ratio k_s/k

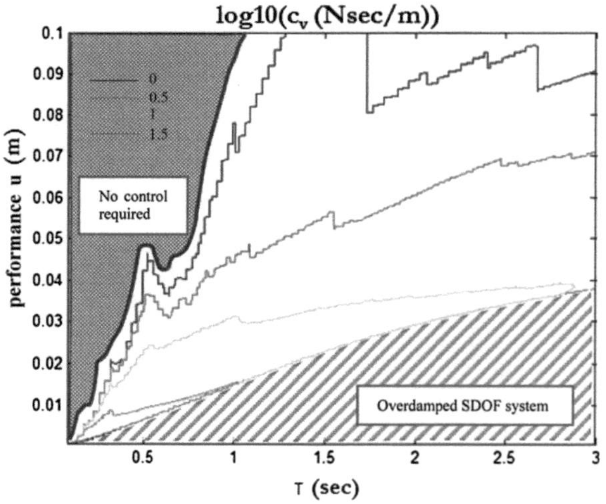

Fig. 6.30 Contour lines of optimal value of c_v related to optimal ratio k_s/k

SDOF system should be overdamped; the red area of the surface, corresponding to very low periods and bounded by the contour line with value 1.0, represents the target displacements can be achieved without any dissipation (passive control) ($\xi = \xi_s = 2\,\%$). It follows that, therefore, in the remaining part of the surface, the integrated design has been developed.

It is deduced that for fixed ratios of cost, the economically optimal combination of the design variables depends on the period of the equivalent single-degree-of-freedom integrated system and the desired performance.

In fact, with reference to cases 1, 2, 3 and 4, as shown in Figs. 6.19, 6.25, 6.31 and 6.37, for periods longer than 0.5 s and for high performance (small displacements) it is always cheaper to use viscoelastic resources and have a minimum

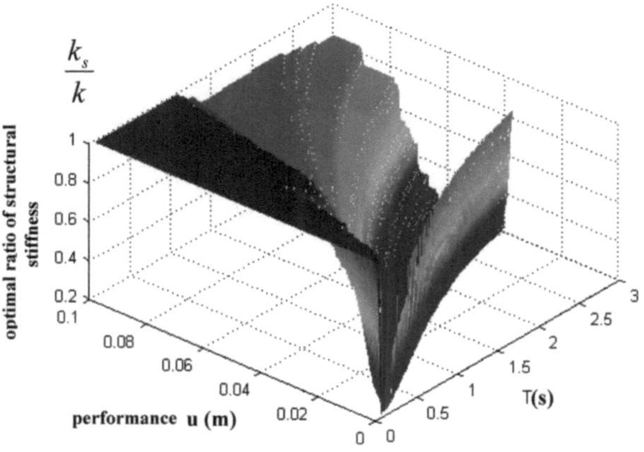

Fig. 6.31 Optimal ratio k_s/k

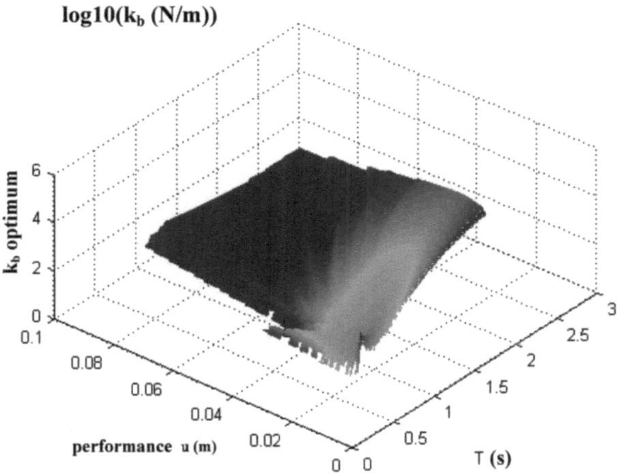

Fig. 6.32 Optimal value of k_b related to optimal ratio k_s/k

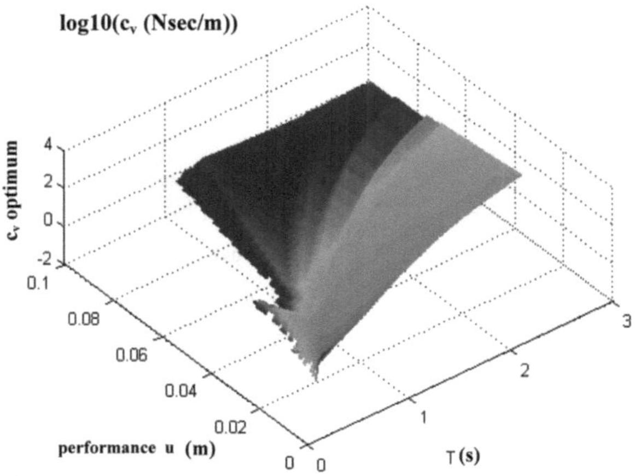

Fig. 6.33 Optimal value of c_v related to optimal ratio k_s/k

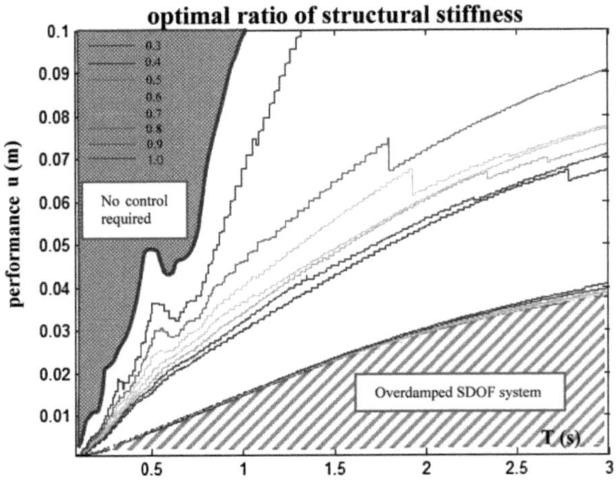

Fig. 6.34 Contour lines of optimal ratio k_s/k

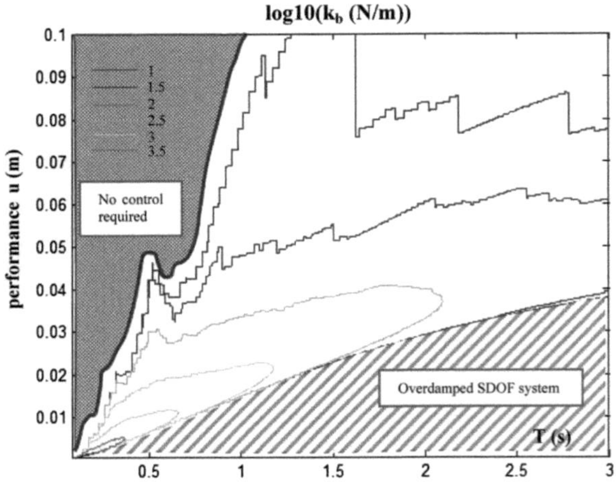

Fig. 6.35 Contour lines of optimal value of k_b related to optimal ratio k_s/k

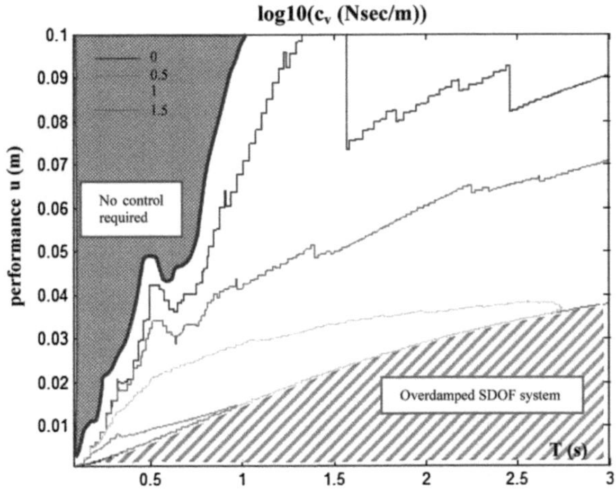

Fig. 6.36 Contour lines of optimal value of c_v related to optimal ratio k_s/k

Fig. 6.37 Optimal ratio k_s/k

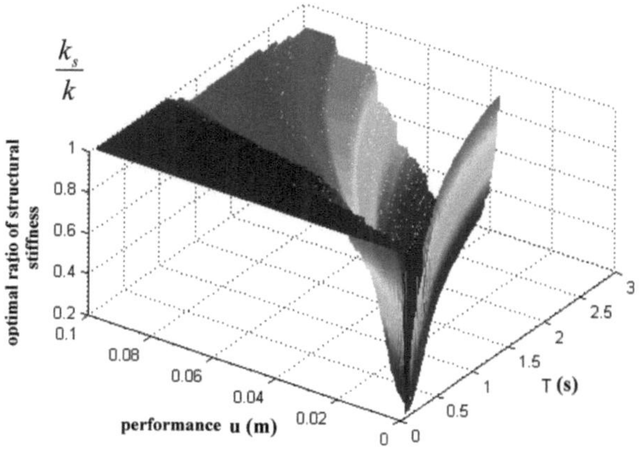

Fig. 6.38 Optimal value of k_b related to optimal ratio k_s/k

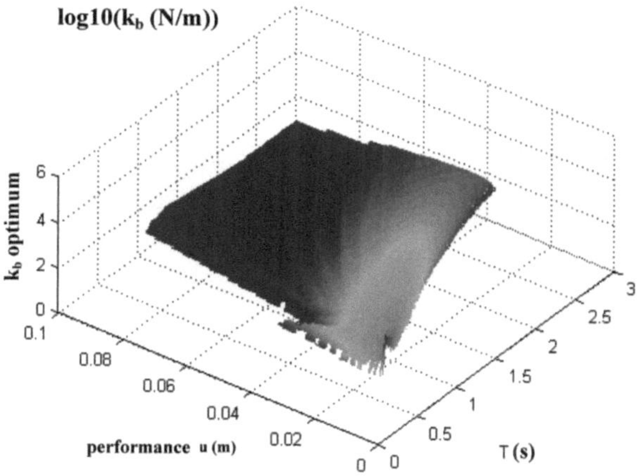

Fig. 6.39 Optimal value of c_v related to optimal ratio k_s/k

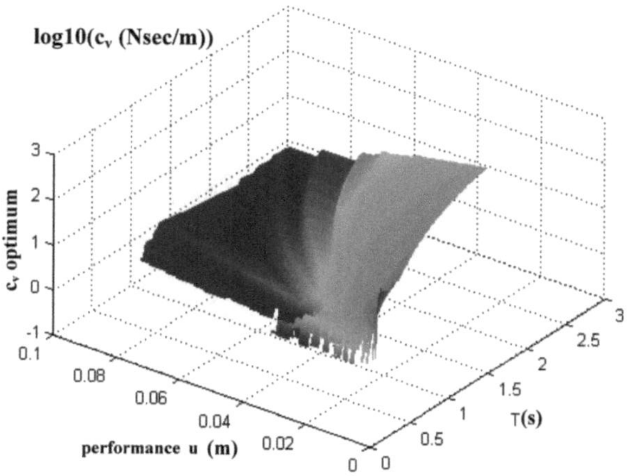

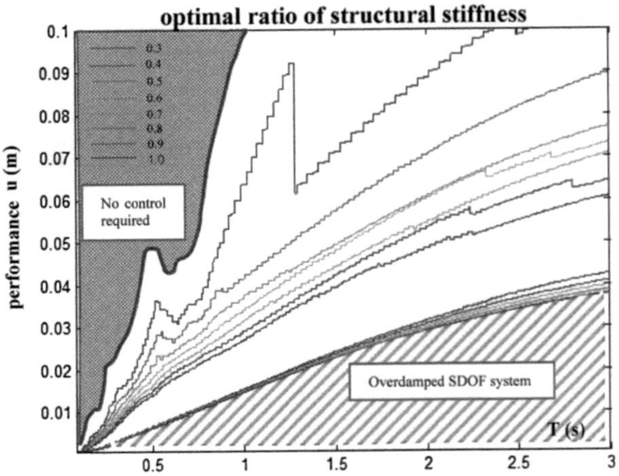

Fig. 6.40 Contour lines of optimal ratio k_s/k

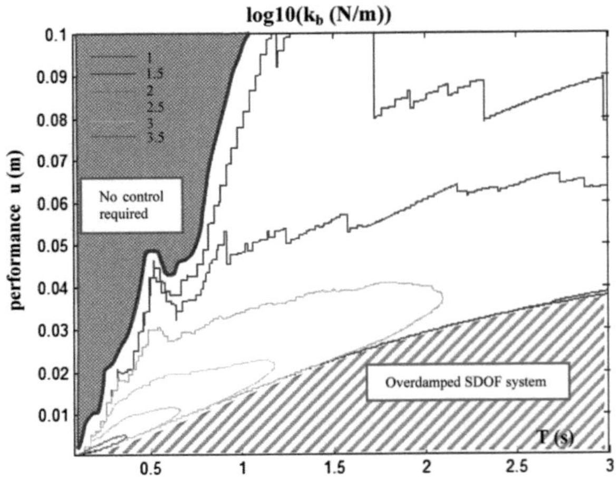

Fig. 6.41 Contour lines of optimal value of k_b related to optimal ratio k_s/k

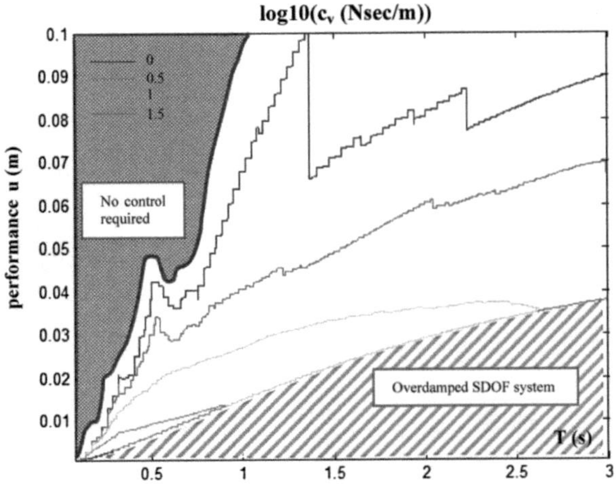

Fig. 6.42 Contour lines of optimal value of c_v related to optimal ratio k_s/k

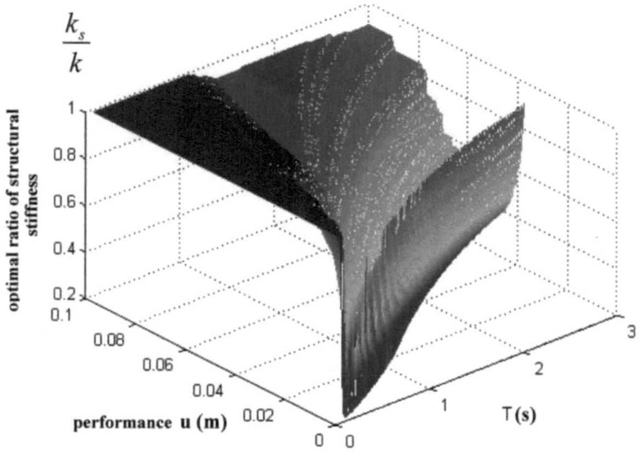

Fig. 6.43 Optimal ratio k_s/k

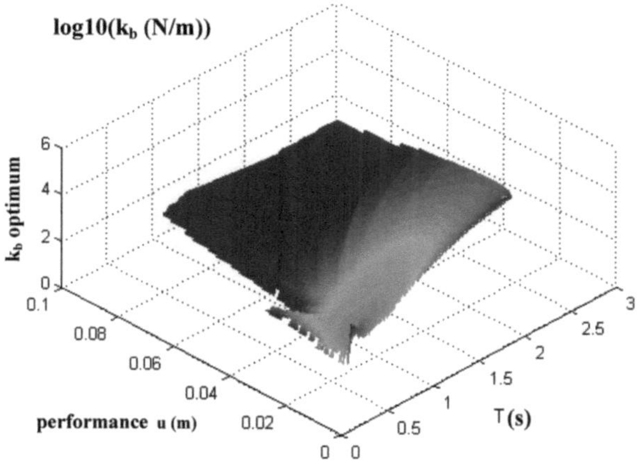

Fig. 6.44 Optimal value of k_b related to optimal ratio k_s/k

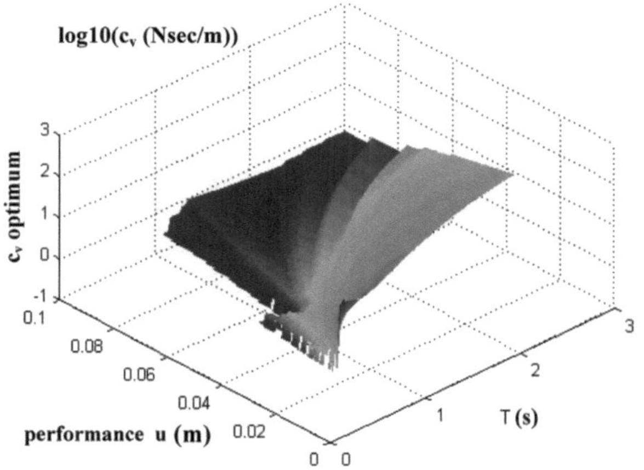

Fig. 6.45 Optimal value of c_v related to optimal ratio k_s/k

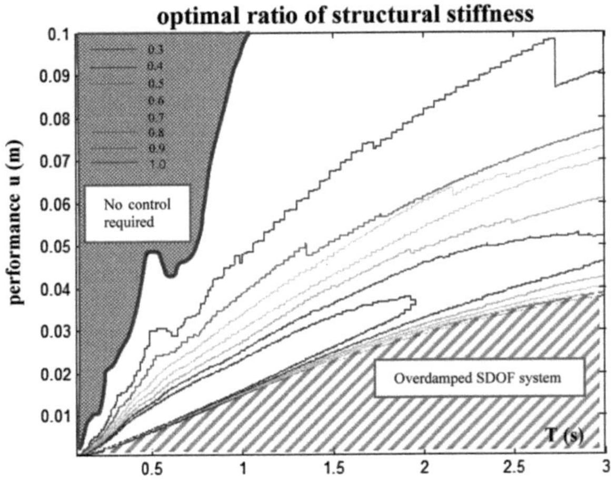

Fig. 6.46 Contour lines of optimal ratio k_s/k

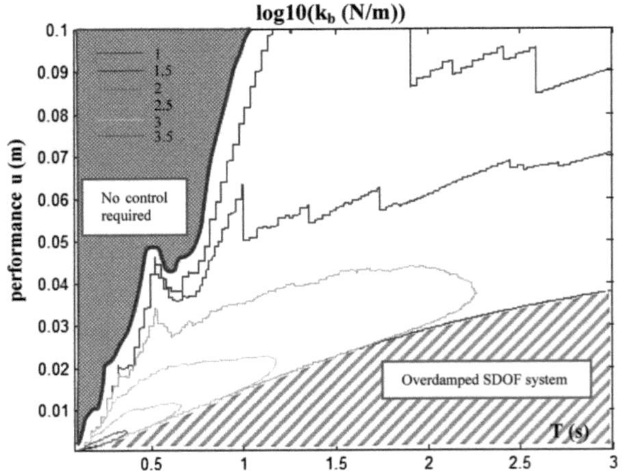

Fig. 6.47 Contour lines of optimal value of k_b related to optimal ratio k_s/k

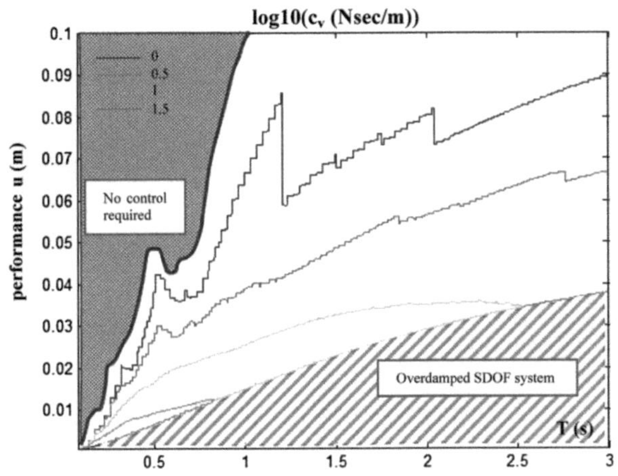

Fig. 6.48 Contour lines of optimal value of c_v related to optimal ratio k_s/k

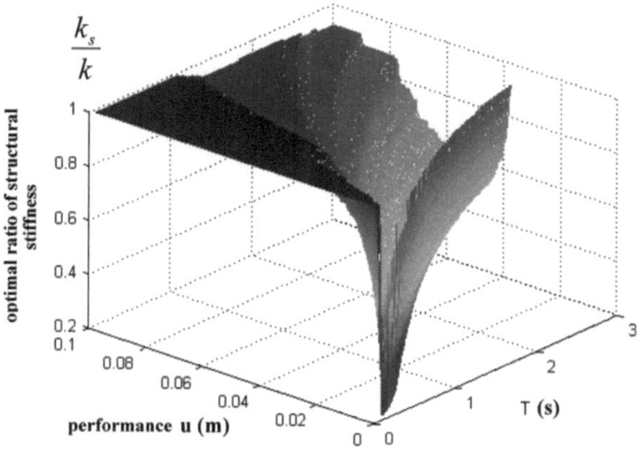

Fig. 6.49 Optimal ratio k_s/k

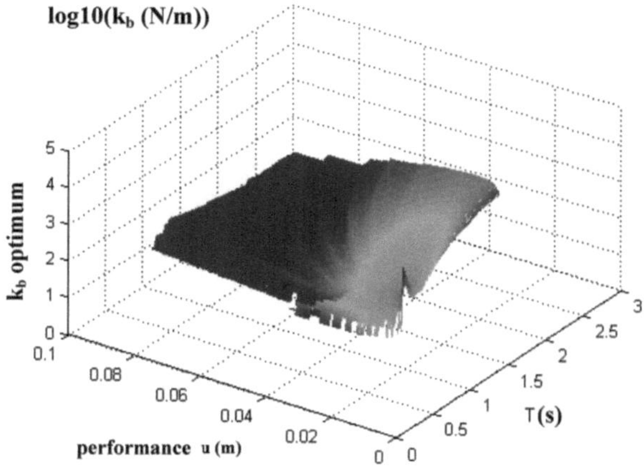

Fig. 6.50 Optimal value of k_b related to optimal ratio k_s/k

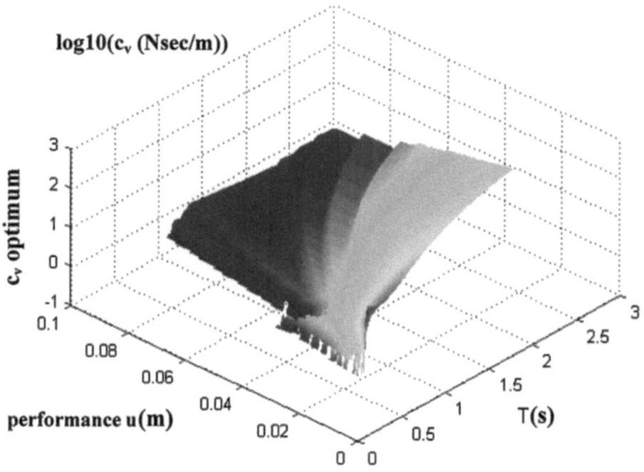

Fig. 6.51 Optimal value of c_v related to optimal ratio k_s/k

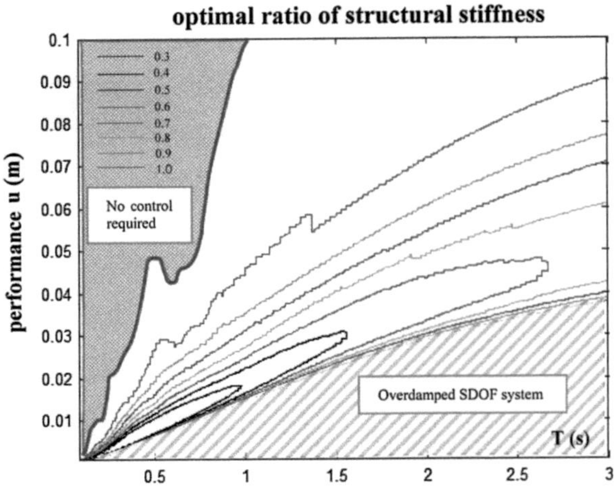

Fig. 6.52 Contour lines of optimal ratio k_s/k

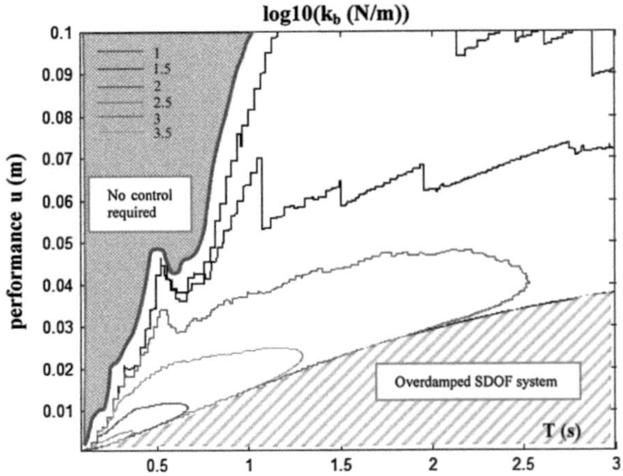

Fig. 6.53 Contour lines of optimal value of k_b related to optimal ratio k_s/k

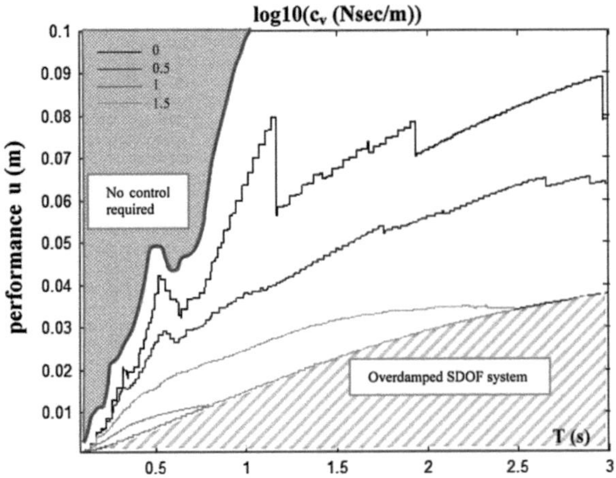

Fig. 6.54 Contour lines of optimal value of c_v related to optimal ratio k_s/k

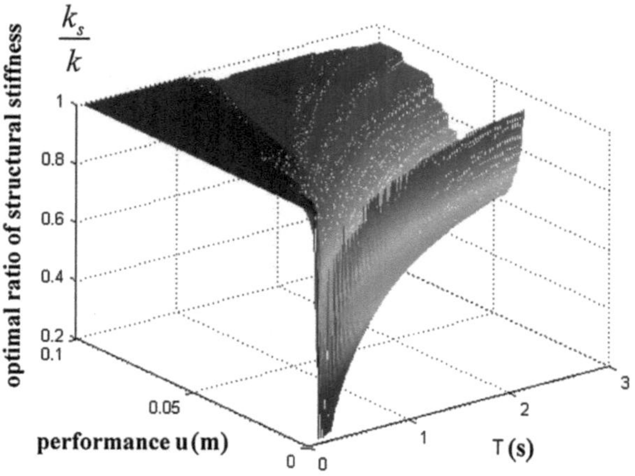

Fig. 6.55 Optimal ratio k_s/k

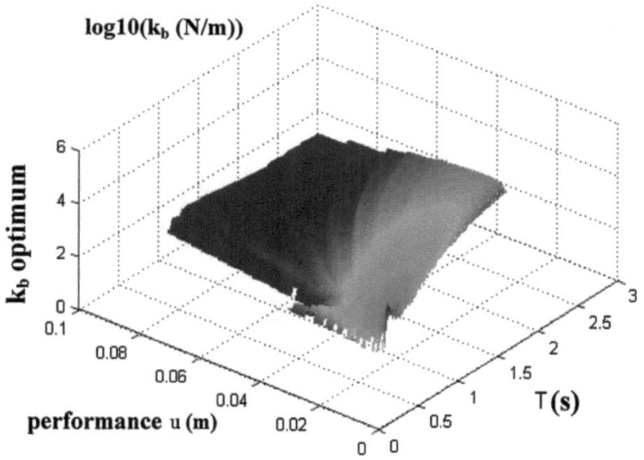

Fig. 6.56 Optimal value of k_b related to optimal ratio k_s/k

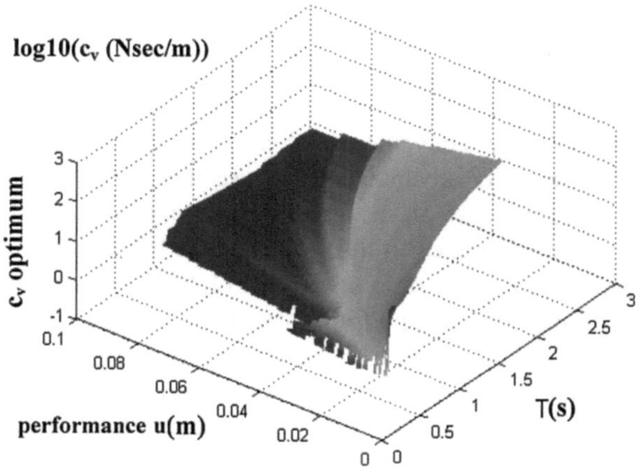

Fig. 6.57 Optimal value of c_v related to optimal ratio k_s/k

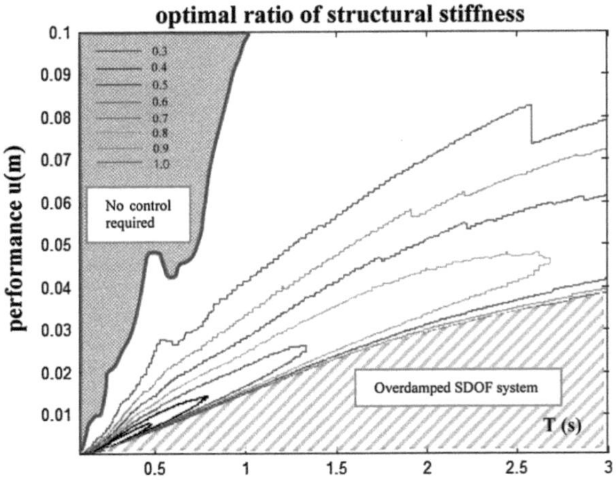

Fig. 6.58 Contour lines of optimal ratio k_s/k

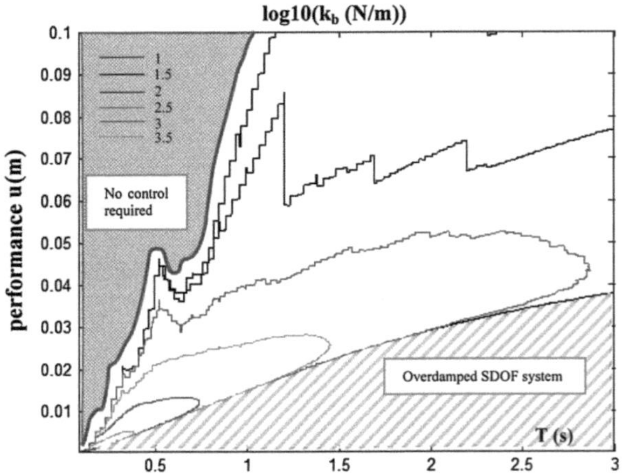

Fig. 6.59 Contour lines of optimal value of k_b related to optimal ratio k_s/k

value of lateral stiffness of the structural system necessary for supporting the gravity loads: k_s/k is equal to 0.2. In the case of low performance (high displacements), having only the lateral stiffness of the structural system is economically more convenient. With reference to periods of less than 0.5 s, it is economically advantageous to use the elastic resources of the structural system also in the case of very high performance.

With reference to the cases 5, 6 and 7, as shown in Figs. 6.43, 6.49 and 6.55, since the relative ratio of the cost C_v/C_s is higher, the optimal design consists, as it is intuitively expected, of an increasing use of elastic resources of the structural system by reducing the viscoelastic resources of the dissipative system.

Figure 6.61, obtained by overlapping, for example, the contour lines of the cases 2 and 4, illustrated in Figs. 6.28 and 6.40, shows the incidence of the relative ratio of cost C_v/C_s on the economically optimal choice: it is clear that, compared to case 2, in case 4, the surface with a high value of lateral stiffness of the

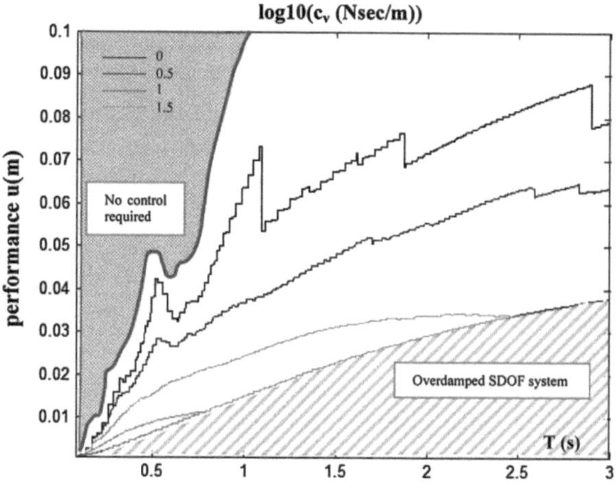

Fig. 6.60 Contour lines of optimal value of c_v related to optimal ratio k_s/k

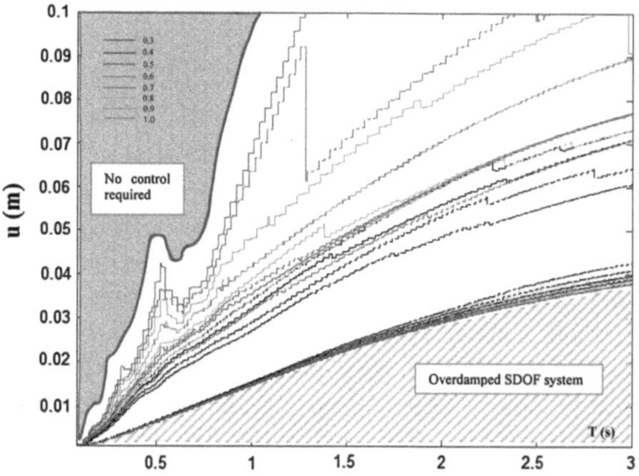

Fig. 6.61 Overlapping the contour lines of optimal ratio k_s/k related to case 2 "—" and case 4 "- -"

structural system is much broader and extends to low performances (high displacements); only with increasing the period and for smaller displacements, is there a trend to exploit the viscoelastic resources without which the desired performance could not be achieved; with decreasing performance (high displacements), it is convenient to have almost exclusively the elastic resources of the structural system.

For each cost ratio, two design values, respectively, representing the optimal values of k_b and c_v, correspond to each design value of the optimal stiffness ratio k_s/k. For example, from Figs. 6.29 and 6.30 it is possible to deduce that a low period structural system demands high value of k_b and less high value of c_v in order to achieve small displacement by developing the design damping ratio ξ. These values of k_b, c_v and ξ lead to a low value of the product $\tau\omega_D$ improving the effectiveness of the viscoelastic dissipative bracing system trough a high value of $c'(\omega_D)$. With reference to high period structural systems and less small displacements, the values of k_b and c_v are less high since ω_D decreases. As for high displacements, this trend is more evident since the required viscoelastic resources are lesser.

From the above-mentioned discussion and critical analysis of the results and figures relating to the proposed parametric study, it can be concluded that the choice of the economically optimal combination of design variables depends on the values of the cost ratios, the natural vibration period and the expected seismic design displacement. For relatively low cost ratios, the supplemental viscoelastic energy dissipation can represent a very important design component to be employed in order to reduce the dynamic response of the integrated system and, thus, any damage.

Finally, it is worth noting that the above results with corresponding curves are extensible also to the case of viscoelastic devices in which the properties of the viscous damper-brace component k_{vb} and c_{vb} correspond, respectively, to k_{vb} and c_{vb} of viscoelastic damper-brace component as shown in Eqs. (4.30) and (4.31).

6.5 Cost Comparison

With reference to an equivalent SDOF system, in this section, the normalized cost of an optimal integrated structural system, equipped with viscoelastic dissipative bracing system, is compared with the normalized ones of both optimal unbraced and braced structures for fixed seismic design performance in order to verify the cost-effectiveness in designing an integrated system compared to conventional seismic-resistant systems which achieve an expected performance requirement.

This comparison was made by considering all the pairs of relative cost ratios (Table 5.4) to evaluate the optimal normalized cost of the integrated system, in accordance with Eq. (5.9), while the normalized cost of the optimal unbraced structure was evaluated as indicated by Eq. (5.10). Finally, the normalized cost of the optimal braced structure has been evaluated as expressed by Eq. (5.11), by always considering $\vartheta = \pi/4$ and a unit mass.

Subsequently, moreover, the above-mentioned comparison was carried out by also considering the natural vibration period as a design variable and, in a second phase, a constraint on the maximum value of the overall viscous damping ratio assumed to be equal to 40 %, too.

With reference to the case 1, $C_b/C_s = 0.5$ and $C_v/C_s = 2$, in Fig. 6.62 the value, in base-e logarithmic scale, of the normalized cost of the optimal integrated structural system is represented for different values of the period T (i.e., undamped natural frequency ω), and seismic design performance u and, in Fig. 6.63, the corresponding contour lines are represented.

From Fig. 6.63, it is possible to deduce that the dashed area of the diagram relative to very small displacements is the area in which the oscillator is over-damped or critically damped. This area has not been taken into account in the analyses. The portion of the surface related to the very low periods is the area in

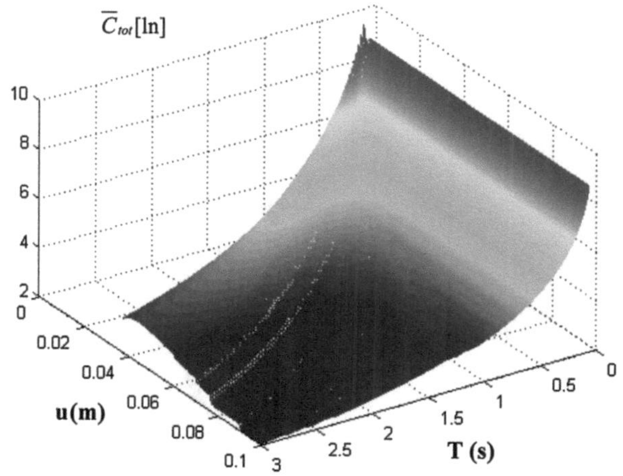

Fig. 6.62 Total normalized cost of the optimal integrated structural system for different values of period and performance

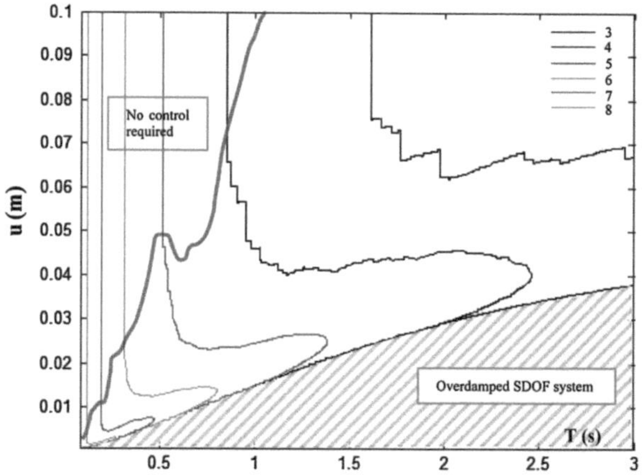

Fig. 6.63 Contour lines of total normalized cost of the optimal integrated structural system for different values of period and performance

which there is $\xi = \xi_s = 2\,\%$ or rather no passive control is required and, therefore, the remaining part of the surface was interested by the optimal integrated design. This consideration is valid for all the cases from 1 to 7.

By also considering the elastic structural period of the integrated system as design variable, the periods, for which the normalized cost of the integrated structural system is minimum, are shown in Fig. 6.64 for different design displacements values.

From Fig. 6.64, it is clear that, since the relative ratio C_v/C_s of this case is low, there is a trend to choose integrated structural systems having a high period with high viscoelastic damping resources resulting the most economically convenient. The trend is also justified by the fact that as the period T increases and, thus, the natural frequency ω decreases, the efficacy of the reduction factor of the static viscosity coefficient increases, as shown in Fig. 4.12. This trend is characterised by jump discontinuities indicating the convenience to choose structures with more stiffness and less viscoelastic damping resources.

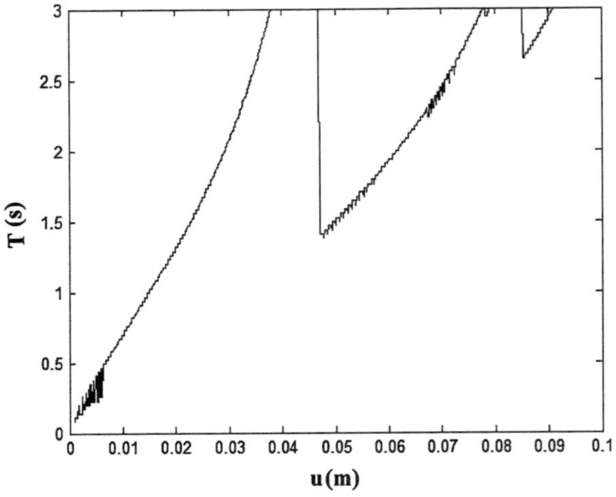

Fig. 6.64 Economically optimal period for different performance values

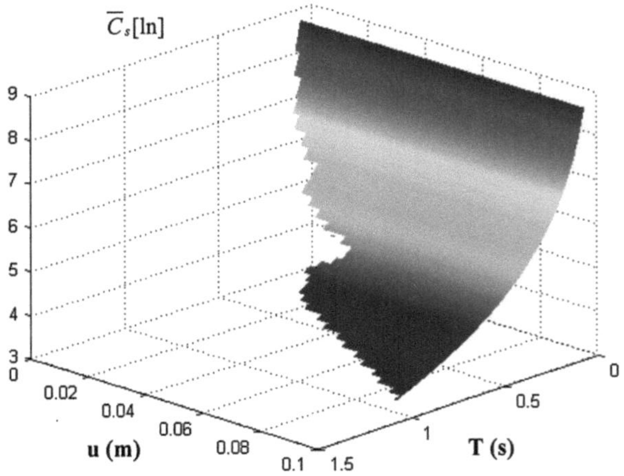

Fig. 6.65 Optimal normalized cost of the unbraced structural system for different values of period and performance

In order to develop an economic comparison between the integrated design solution with two other conventional design solutions, unbraced or braced structural systems, the costs of economically optimal configurations of the two conventional design solutions were also evaluated.

In Fig. 6.65, the value, in base-e logarithmic scale, of the normalized cost of the optimal unbraced structural system is shown for different values of the period T (i.e., undamped natural frequency ω), and seismic performance u and in Fig. 6.66 the corresponding contour lines are represented.

In Fig. 6.67, the value, in base-e logarithmic scale, of the normalized cost of the optimal braced structural system is shown for different values of the period T (i.e., undamped natural frequency ω), and seismic design displacement u and, in Fig. 6.68, the corresponding contour lines are represented. The above-mentioned cost has been assessed by giving 80 % of the total lateral stiffness to the bracing

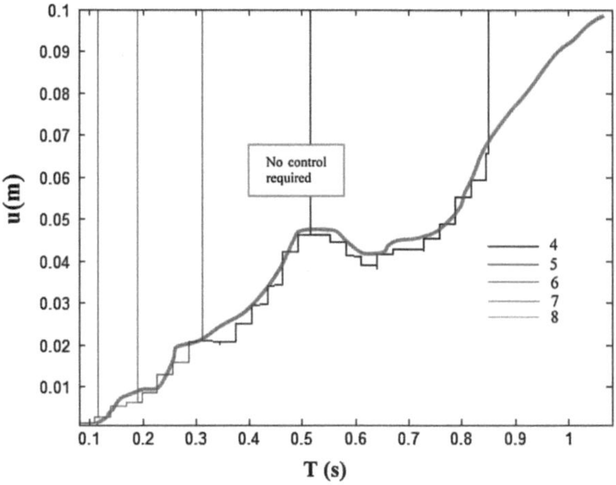

Fig. 6.66 Contour lines of optimal normalized cost of the unbraced structural system for different values of period and performance

system and 20 %, which is the minimum imposed and discussed in Sects. 5.3 and 5.4, to the lateral stiffness of the structural system. The minimum value of lateral stiffness of the structural system is chosen since, on the basis of the relative ratio of cost $C_b/C_s = 0.5$, this solution always provides the optimal cost.

From Figs. 6.65, 6.66, 6.67 and 6.68, it is possible to observe that only the area in which $\xi = 2\,\%$ has been considered because there is no need of passive dissipation to achieve the target performance. With reference to the solutions of both unbraced and braced structures, the economic optimal solution, by considering the vibration period as a design variable, is achievable by regarding the highest period, i.e. the system with the lowest stiffness. These considerations related to the design solution both of the unbraced and braced structures are, obviously, valid for all the cases from 1 to 7.

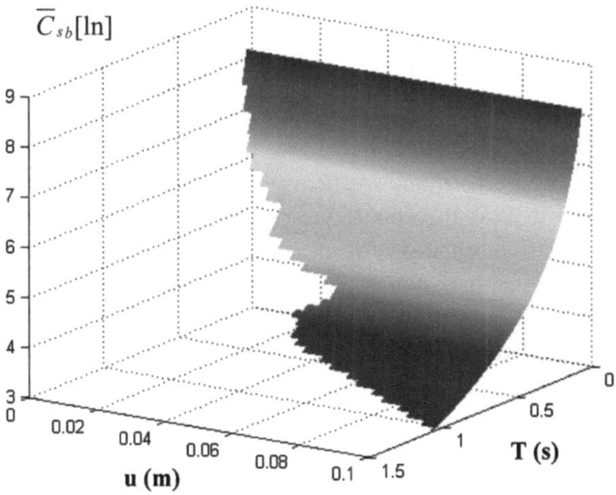

Fig. 6.67 Optimal normalized cost of the braced structural system for different values of period and performance

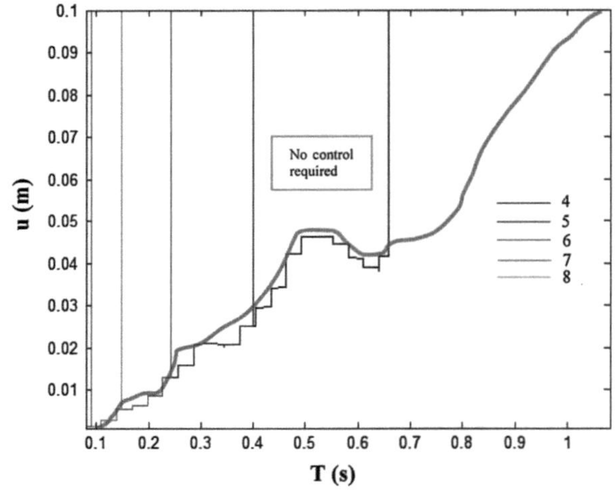

Fig. 6.68 Contour lines of optimal normalized cost of the braced structural system for different values of period and performance

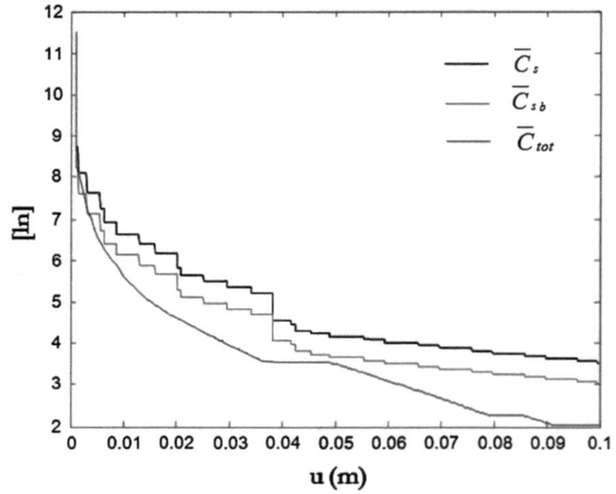

Fig. 6.69 Comparison between normalized costs related to the optimal design solutions considered for different performance values

In Fig. 6.69, therefore, the minimum values of the normalized cost of the integrated structural system $\overline{C}_{tot}$, the normalized ones of both unbraced $\overline{C}_s$ and braced $\overline{C}_{sb}$ structures relative to the optimal periods are represented for different seismic performance values.

Note that the integrated system, consisting of the structural and viscoelastic dissipative bracing systems, is the cheapest for all the seismic design displacements.

In a second phase, a constraint on the maximum value of the supplemental damping ratio has been imposed that's the supplemental damping produced by the viscoelastic dissipative system may not exceed 38 % such that the integrated system presents an overall viscous damping ratio equal to 40 %. The periods, for which the normalized cost of the integrated system is minimum, are represented in

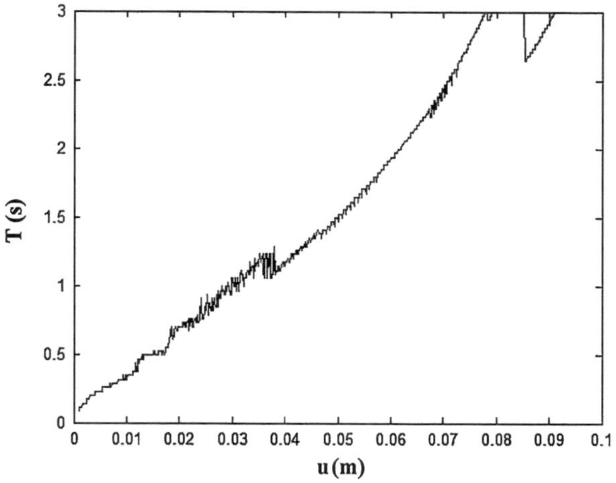

Fig. 6.70 Economically optimal period for different values of performance considering the upper limit on the damping ratio equal to 40 %

Fig. 6.70 for different performance values in the presence of the upper limit on the damping ratio.

From Fig. 6.70, it can be highlighted that, unlike the situation without constraint on the damping ratio, there is a more continuous and gradual trend to choose integrated structural systems having a high period with high viscoelastic damping resources resulting the most economically convenient. This trend, as already mentioned, is also justified by the fact that as the period T increases and, so, the natural frequency ω decreases, the efficacy of the reduction factor of the static viscosity coefficient increases, as is shown in Fig. 4.12. In Fig. 6.71, the damping value, limited superiorly to 40 %, related to each period of the economically optimal integrated system relating to each seismic performance is shown.

Finally, in Fig. 6.72, the minimum values of the normalized cost of the integrated structural system $\overline{C}_{tot}$, having a damping ratio at most equal to 40 %, the

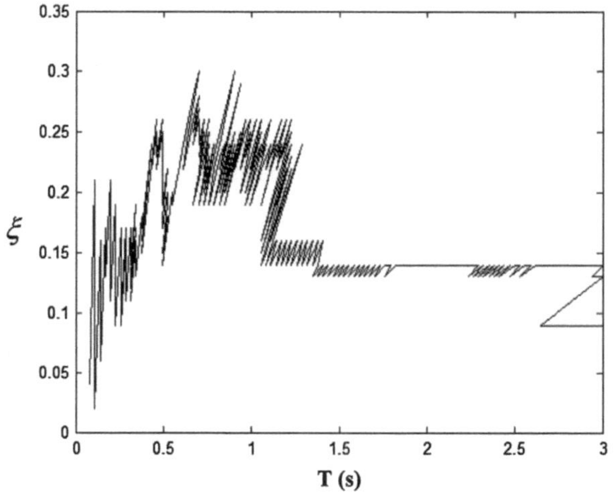

Fig. 6.71 Damping value, limited superiorly to 40 %, related to each optimal period and each performance

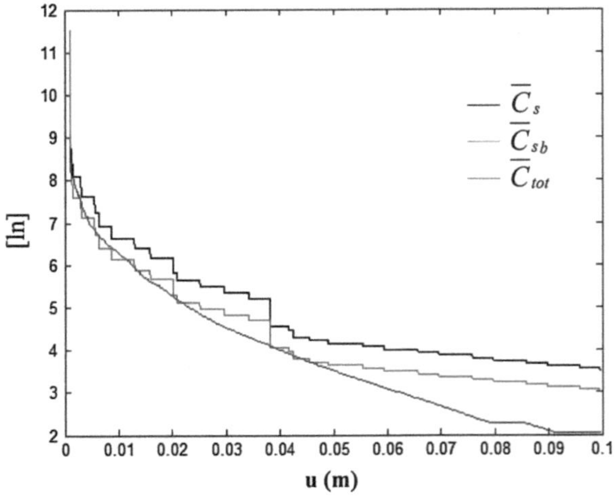

Fig. 6.72 Comparison between normalized costs related to the optimal design solutions considered for different values of performance with the upper limit on the damping ratio equal to 40 %

normalized ones of both unbraced $\overline{C}_s$ and braced $\overline{C}_{sb}$ structures relative to the optimal periods are represented.

Similarly, with reference to case 2, $C_b/C_s = 0.5$ and $C_v/C_s = 3$, the value, in base-e logarithmic scale, of the normalized cost of the optimal integrated structural system for different values of the period T (i.e., undamped natural frequency ω), and seismic design performance u is represented in Fig. 6.73 and, in Fig. 6.74, the corresponding contour lines are represented.

The periods, for which the normalized cost of the integrated structural system is minimum, are represented in Fig. 6.75 for different seismic design displacement values.

With regard to Fig. 6.75 the considerations discussed relating to the previous case are the same.

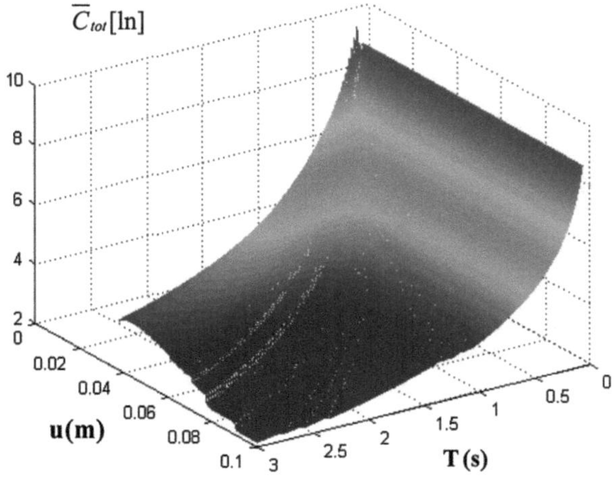

Fig. 6.73 Total normalized cost of the optimal integrated structural system for different values of period and performance

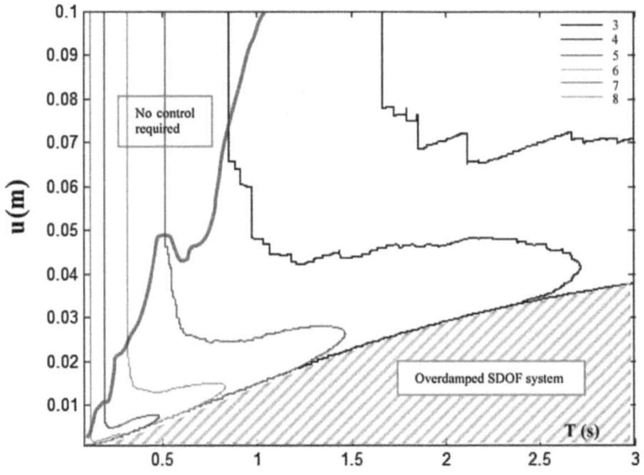

Fig. 6.74 Contour lines of total normalized cost of the optimal integrated structural system for different values of period and performance

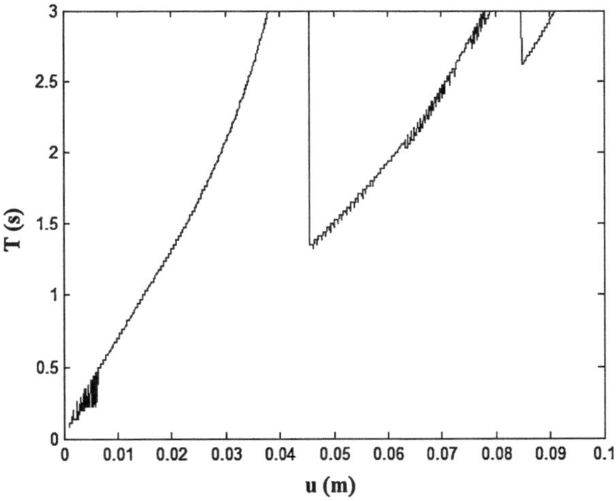

Fig. 6.75 Economically optimal period for different performance values

With reference both to the normalized cost of the optimal unbraced structural system as well as the normalized cost of the optimal braced structural system evaluated for different values of the period T (i.e., undamped natural frequency ω) and seismic performance u, Figs. 6.65, 6.66, 6.67 and 6.68 remain unchanged.

In Fig. 6.76, the minimum values of the normalized cost of the integrated structural system $\overline{C}_{tot}$, the normalized ones of both unbraced $\overline{C}_s$ and braced $\overline{C}_{sb}$ structures relative to the optimal periods are represented.

It can be noted that, even in this case, the integrated system, constituted by the structural system and viscoelastic dissipative bracing system, is always cheaper than the other solutions for every performance considered.

By imposing the constraint on the maximum value of the overall damping ratio equal to 40 %, the periods, for which the normalized cost of the integrated structural system is minimum, are represented in Fig. 6.77 for different seismic performance values.

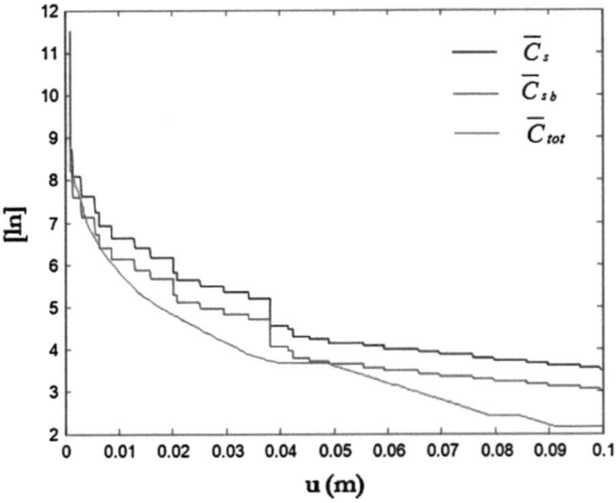

Fig. 6.76 Comparison between normalized costs related to the optimal design solutions considered for different performance values

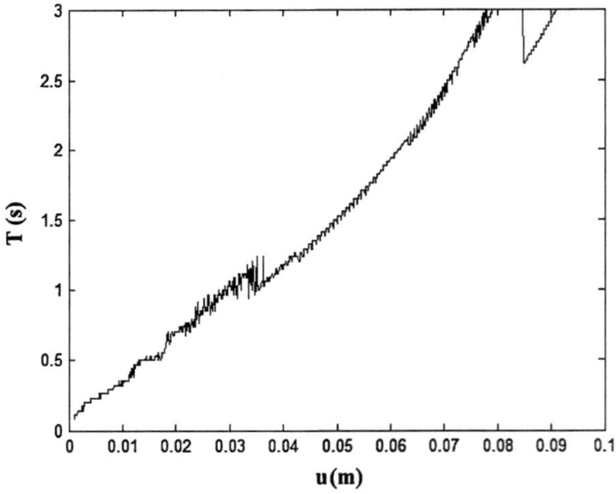

Fig. 6.77 Economically optimal period for different values of performance considering the upper limit on the damping ratio equal to 40 %

In Fig. 6.77, a trend similar to that discussed in the previous case can be observed.

In Fig. 6.78, the damping value, limited superiorly to 40 %, related to each period of the economically optimal integrated system relatively to each seismic performance is shown.

Finally, in Fig. 6.79, the minimum values of the normalized cost of the integrated structural system $\overline{C}_{tot}$, having a damping ratio at most equal to 40 %, the normalized ones of both unbraced $\overline{C}_s$ and braced $\overline{C}_{sb}$ structures relative to the optimal periods are represented.

Similarly, with reference to case 3, $C_b/C_s = 0.5$ and $C_v/C_s = 5$, the value, in base-e logarithmic scale, of the normalized cost of the optimal integrated structural system for different values of the period T (i.e., undamped natural frequency ω),

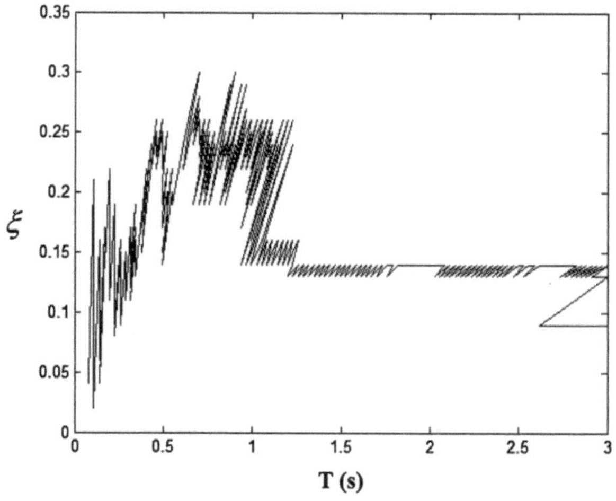

Fig. 6.78 Damping value, limited superiorly to 40 %, related to each optimal period and each performance

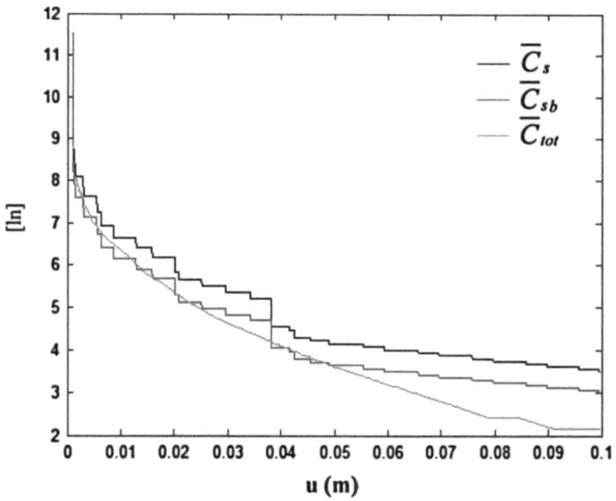

Fig. 6.79 Comparison between normalized costs related to the optimal design solutions considered for different values of performance with the upper limit on the damping ratio equal to 40 %

and seismic design displacement u is represented in Fig. 6.80 and, in Fig. 6.81, the corresponding contour lines are represented.

The periods, for which the normalized cost of the integrated structural system is minimum, are represented in Fig. 6.82 for different seismic performance values.

Regarding Fig. 6.82, the considerations discussed relating to the previous case are the same.

With reference both to the normalized cost of the optimal unbraced structural system as well as the normalized cost of the optimal braced structural system evaluated for different values of the period T (i.e., undamped natural frequency ω) and seismic performance u, Figs. 6.65, 6.66, 6.67 and 6.68 remain unchanged.

In Fig. 6.83, the minimum values of the normalized cost of the integrated structural system $\overline{C}_{tot}$, the normalized ones of both unbraced $\overline{C}_s$ and braced $\overline{C}_{sb}$ structures relative to the optimal periods are represented.

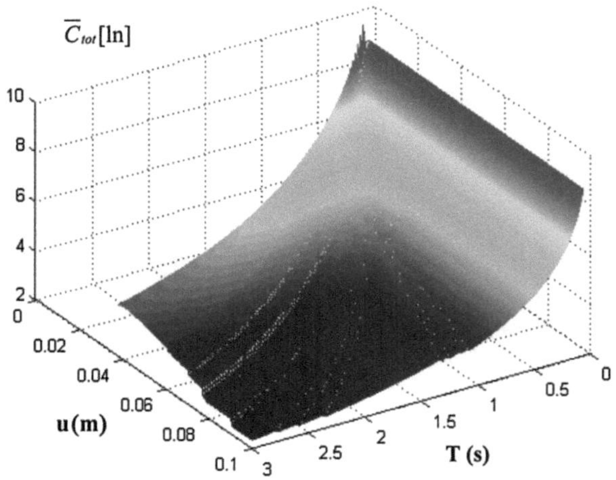

Fig. 6.80 Total normalized cost of the optimal integrated structural system for different values of period and performance

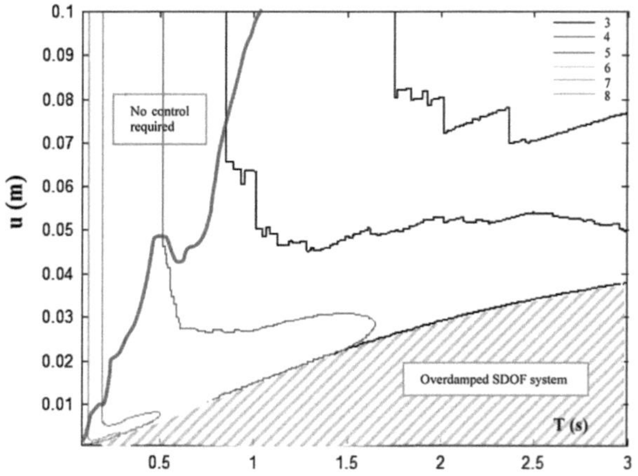

Fig. 6.81 Contour lines of total normalized cost of the optimal integrated structural system for different values of period and performance

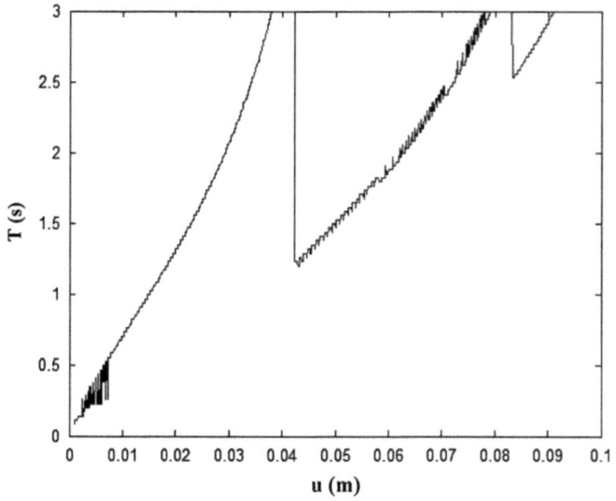

Fig. 6.82 Economically optimal period for different performance values

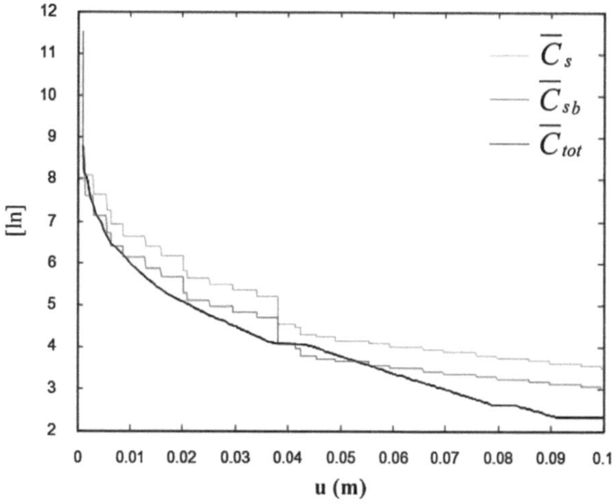

Fig. 6.83 Comparison between normalized costs related to the optimal design solutions considered for different performance values

It can be noted that the integrated system, constituted by the structural system and the viscoelastic dissipative bracing system, is cheaper than the other solutions for almost all seismic design displacements considered.

By imposing the constraint on the maximum value of the overall damping ratio equal to 40 %, the periods, for which the normalized cost of the integrated structural system is minimum, are represented in Fig. 6.84 for different seismic performance values.

In Fig. 6.84, a trend similar to that discussed in the previous case can be observed.

In Fig. 6.85, the damping value, limited superiorly to 40 %, related to each period of the economically optimal integrated system relatively to each seismic performance is shown.

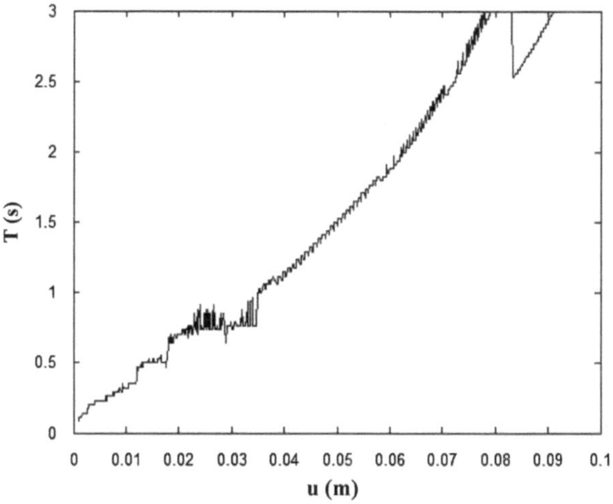

Fig. 6.84 Economically optimal period for different performance values considering the upper limit on the damping ratio equal to 40 %

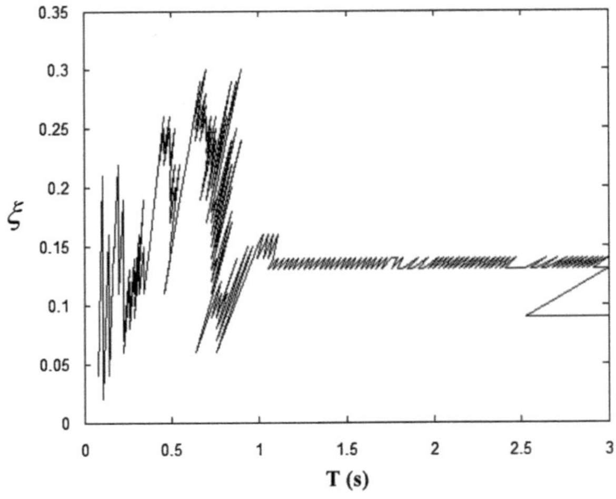

Fig. 6.85 Damping value, limited superiorly to 40 %, related to each optimal period and each performance

Finally, in Fig. 6.86, the minimum values of the normalized cost of the integrated structural system $\overline{C}_{tot}$, having a damping ratio at most equal to 40 %, the normalized ones of both unbraced $\overline{C}_s$ and braced $\overline{C}_{sb}$ structures relative to the optimal periods are represented.

Similarly, with reference to case 4, $C_b/C_s = 0.5$ and $C_v/C_s = 10$, the value, in base-e logarithmic scale, of the normalized cost of the optimal integrated structural system for different values of the period T (i.e., undamped natural frequency ω), and seismic design displacement u is represented in Fig. 6.87 and, in Fig. 6.88, the corresponding contour lines are represented.

The periods, for which the normalized cost of the integrated structural system is minimum, are represented in Fig. 6.89 for different seismic performance values.

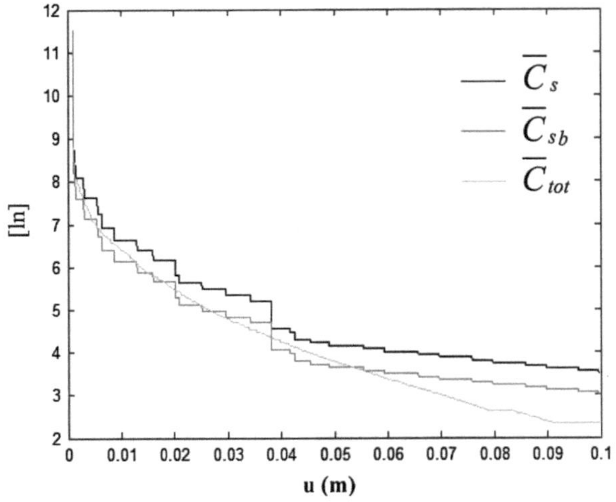

Fig. 6.86 Comparison between normalized costs related to the optimal design solutions considered for different performance values with the upper limit on the damping ratio equal to 40 %

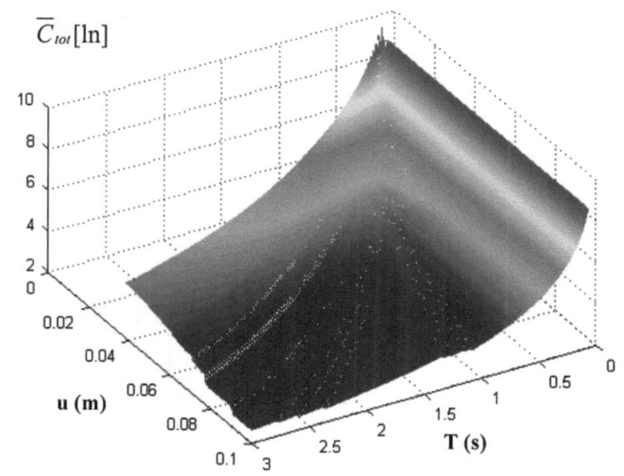

Fig. 6.87 Total normalized cost of the optimal integrated structural system for different values of period and performance

From Fig. 6.89, it is clear that, since the relative ratio C_v/C_s of this case is medium, there is a sufficient gradual trend to choose integrated structural systems having a high period with high viscoelastic damping resources resulting the most economically convenient. The trend is also justified by the fact that as the period T increases and, thus, the natural frequency ω decreases, the efficacy of the reduction factor of the static viscosity coefficient increases, as is shown in Fig. 4.12. This trend is characterised by jump discontinuities indicating the convenience to choose structures with more stiffness and less viscoelastic damping resources.

With reference both to the normalized cost of the optimal unbraced structural system as well as the normalized cost of the optimal braced structural system evaluated for different values of the period T (i.e., undamped natural frequency ω) and target performance u, Figs. 6.65, 6.66, 6.67 and 6.68 remain unchanged.

In Fig. 6.90, the minimum values of the normalized cost of the integrated structural system $\overline{C}_{tot}$, the normalized ones of both unbraced $\overline{C}_s$ and braced $\overline{C}_{sb}$ structures relative to the optimal periods are represented.

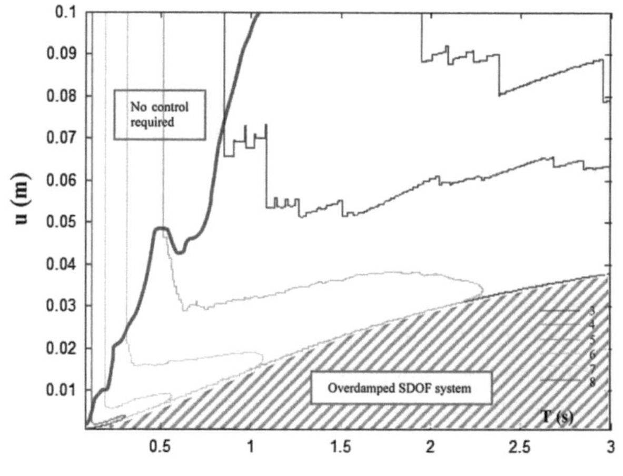

Fig. 6.88 Contour lines of total normalized cost of the optimal integrated structural system for different values of period and performance

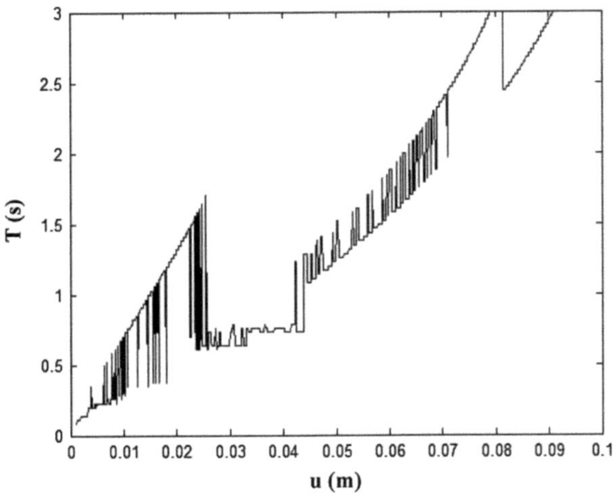

Fig. 6.89 Economically optimal period for different performance values

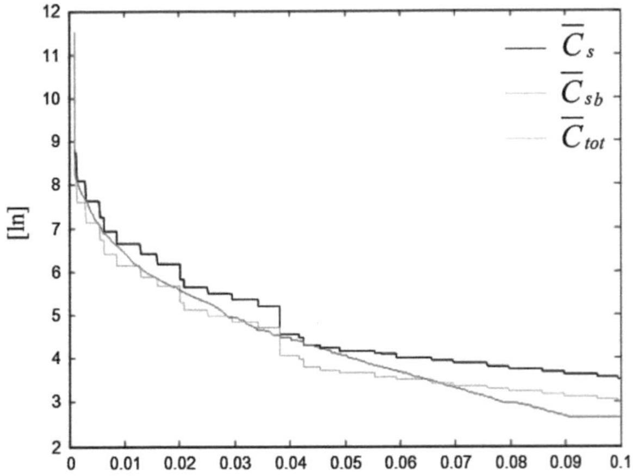

Fig. 6.90 Comparison between normalized costs related to the optimal design solutions considered for different performance values

It can be noted that the integrated system, constituted by the structural system and the viscoelastic dissipative bracing system, is not always cheaper than the other solutions for every seismic performance considered.

By imposing the constraint on the maximum value of the overall damping ratio equal to 40 %, the periods, for which the normalized cost of the integrated structural system is minimum, are represented in Fig. 6.91 for different seismic performance values.

In Fig. 6.91, a more gradual trend with less jump discontinuities can be observed.

In Fig. 6.92, the damping value, limited superiorly to 40 %, related to each period of the economically optimal integrated system relatively to each target performance is shown.

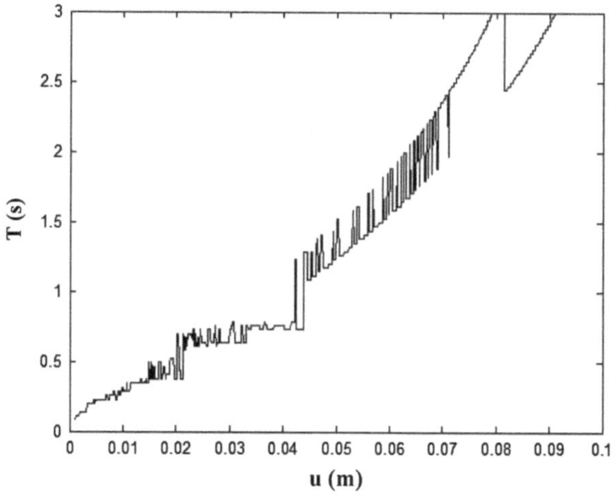

Fig. 6.91 Economically optimal period for different performance values considering the upper limit on the damping ratio equal to 40 %

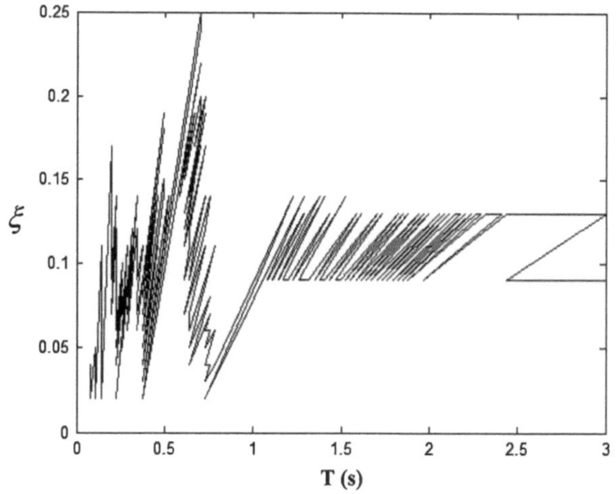

Fig. 6.92 Damping value, limited superiorly to 40 %, related to each optimal period and each performance

Finally, in Fig. 6.93, the minimum values of the normalized cost of the integrated structural system $\overline{C}_{tot}$, having a damping ratio at most equal to 40 %, the normalized ones of both unbraced $\overline{C}_s$ and braced $\overline{C}_{sb}$ structures relative to the optimal periods are represented.

Similarly, with reference to the 5, $C_b/C_s = 0.5$ and $C_v/C_s = 25$, the value, in base-e logarithmic scale, of the normalized cost of the optimal integrated structural system for different values of the period T (i.e., undamped natural frequency ω), and seismic design displacement u is represented in Fig. 6.94 and, in Fig. 6.95, the corresponding contour lines are represented.

The periods, for which the normalized cost of the integrated structural system is minimum, are represented in Fig. 6.96 for different seismic performance values.

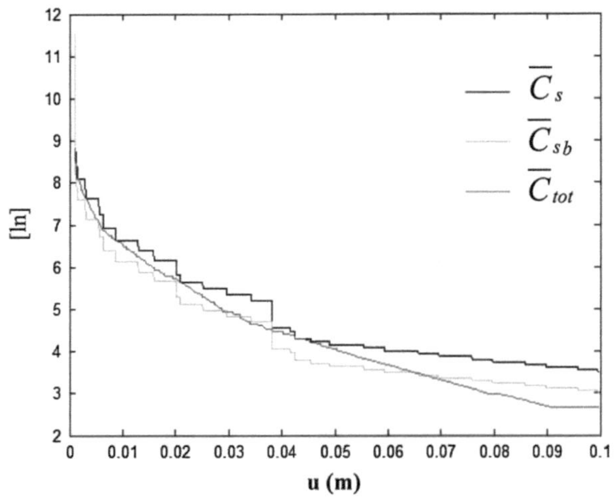

Fig. 6.93 Comparison between normalized costs related to the optimal design solutions considered for different performance values with the upper limit on the damping ratio equal to 40 %

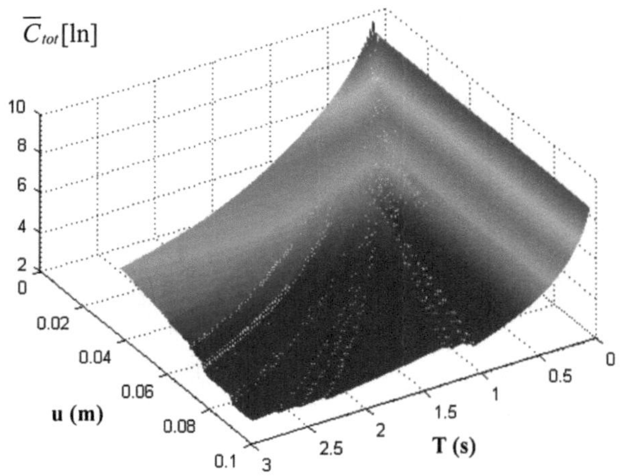

Fig. 6.94 Total normalized cost of the optimal integrated structural system for different values of period and performance

From Fig. 6.96, it is clear that, since the relative ratio C_v/C_s of this case is high, there is a trend to choose integrated structural systems having a less high period with less high viscoelastic damping resources resulting the most economically convenient. This trend is always characterised by jump discontinuities indicating the convenience to choose structures with more stiffness and less viscoelastic damping resources.

With reference both to the normalized cost of the optimal unbraced structural system as well as the normalized cost of the optimal braced structural system evaluated for different values of the period T (i.e., undamped natural frequency ω) and seismic performance u, Figs. 6.65, 6.66, 6.67 and 6.68 remain unchanged.

In Fig. 6.97, the minimum values of the normalized cost of the integrated structural system $\overline{C}_{tot}$, the normalized ones of both unbraced $\overline{C}_s$ and braced $\overline{C}_{sb}$ structures relative to the optimal periods are represented.

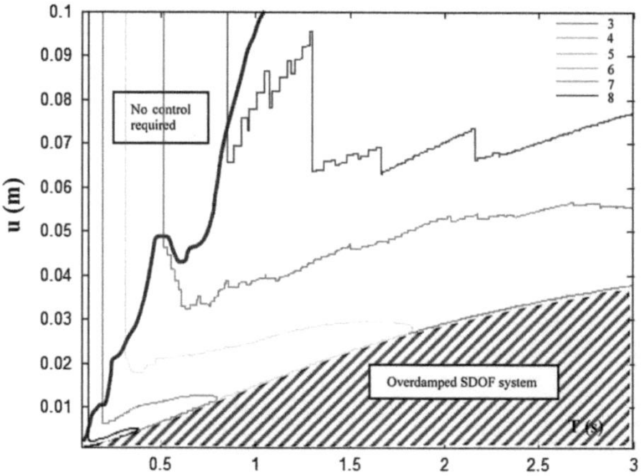

Fig. 6.95 Contour lines of total normalized cost of the optimal integrated structural system for different values of period and performance

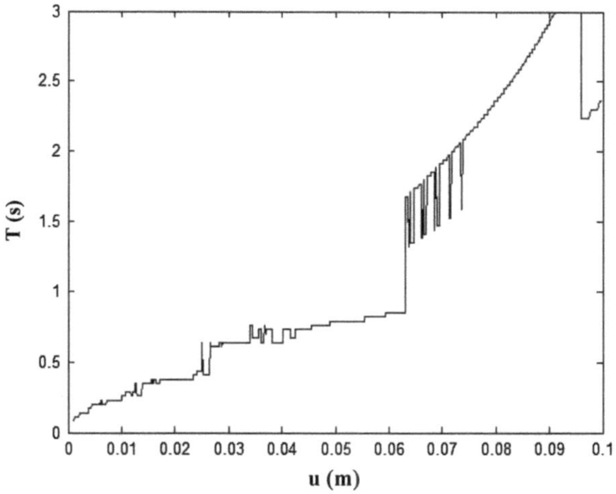

Fig. 6.96 Economically optimal period for different performance values

It can be noted that the integrated system, constituted by the structural system and the viscoelastic dissipative bracing system, is never cheaper than the solution of the braced structural system and it is cheaper than the solution of the unbraced structural system for some seismic design displacements considered.

The constraint on the maximum value of the overall damping ratio, in this case, is respected.

Similarly, with reference to the 6, $C_b/C_s = 0.5$ and $C_v/C_s = 50$, the value, in base-e logarithmic scale, of the normalized cost of the optimal integrated structural system for different values of the period T (i.e., undamped natural frequency ω), and seismic performance u is represented in Fig. 6.98 and, in Fig. 6.99, the corresponding contour lines are represented.

The periods, for which the normalized cost of the integrated structural system is minimum, are represented in Fig. 6.100 for different seismic performance values.

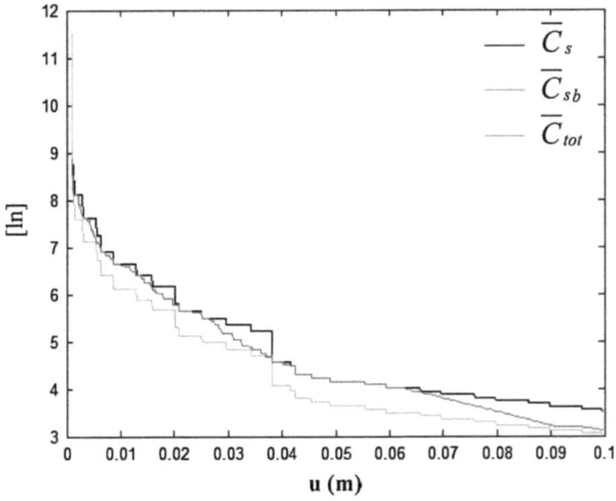

Fig. 6.97 Comparison between normalized costs related to the optimal design solutions considered for different performance values

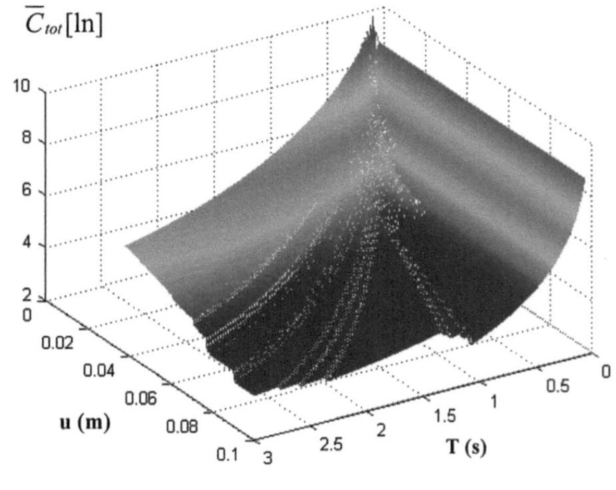

Fig. 6.98 Total normalized cost of the optimal integrated structural system for different values of period and performance

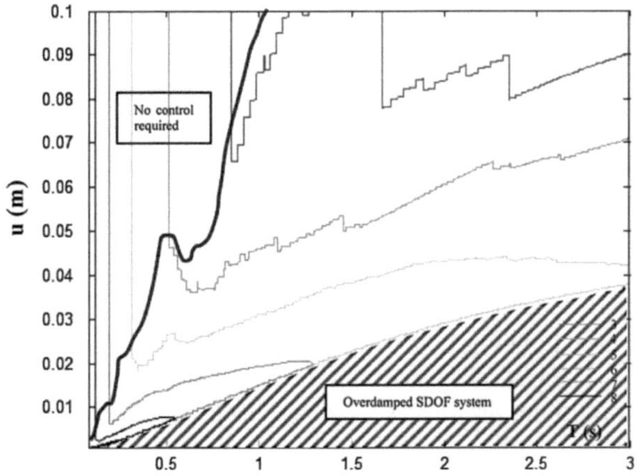

Fig. 6.99 Contour lines of total normalized cost of the optimal integrated structural system for different values of period and performance

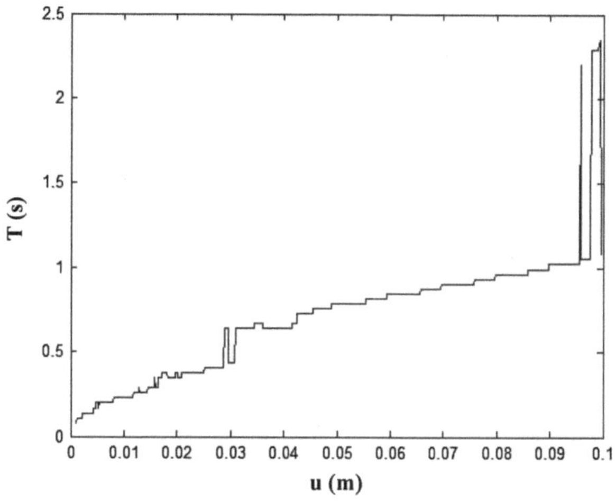

Fig. 6.100 Economically optimal period for different performance values

From Fig. 6.100, it is clear that, since the relative ratio C_v/C_s of this case is high, like in the previous case, there is a trend to choose integrated structural systems having a less high period with less high viscoelastic damping resources resulting the most economically convenient.

With reference both to the normalized cost of the optimal unbraced structural system as well as the normalized cost of the optimal braced structural system evaluated for different values of the period T (i.e., undamped natural frequency ω) and seismic design displacement u, Figs. 6.65, 6.66, 6.67 and 6.68 remain unchanged.

In Fig. 6.101, the minimum values of the normalized cost of the integrated structural system $\overline{C}_{tot}$, the normalized ones of both unbraced $\overline{C}_s$ and braced $\overline{C}_{sb}$ structures relative to the optimal periods are represented.

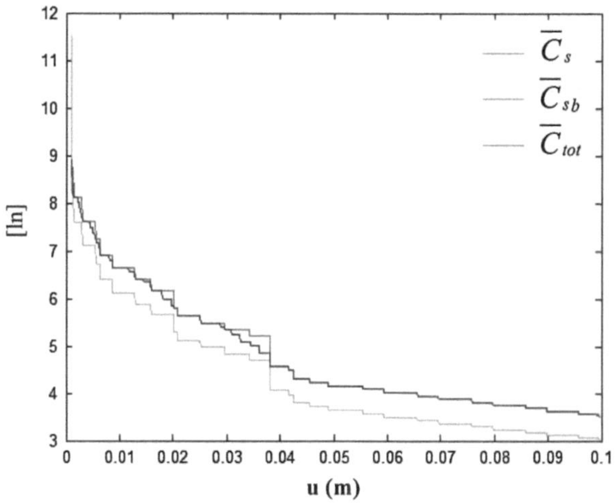

Fig. 6.101 Comparison between normalized costs related to the optimal design solutions considered for different performance values

It can be observed that the integrated system, constituted by the structural system and the viscoelastic dissipative bracing system, is never cheaper than the solution of the braced structural system and it is cheaper than the solution of the unbraced structural system for a few of the seismic design displacements considered.

The constraint on the maximum value of the overall damping ratio, also in this case, is respected.

Similarly, with reference to the 7, $C_b/C_s = 0.5$ and $C_v/C_s = 100$, the value, in base-e logarithmic scale, of the normalized cost of the optimal integrated structural system for different values of the period T (i.e., undamped natural frequency ω), and target performance u is represented in Fig. 6.102 and, in Fig. 6.103, the corresponding contour lines are represented.

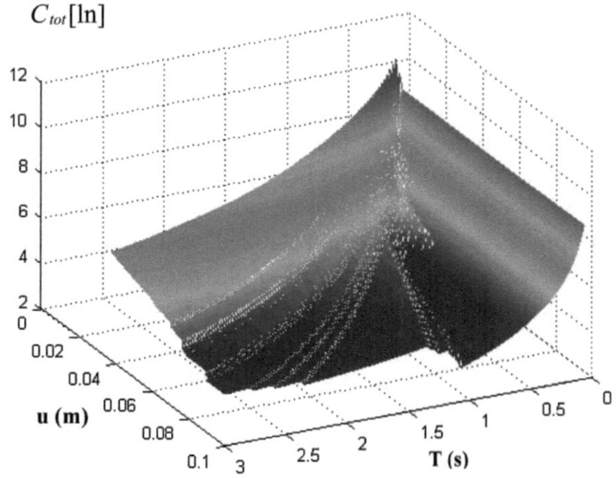

Fig. 6.102 Total normalized cost of the optimal integrated structural system for different values of period and performance

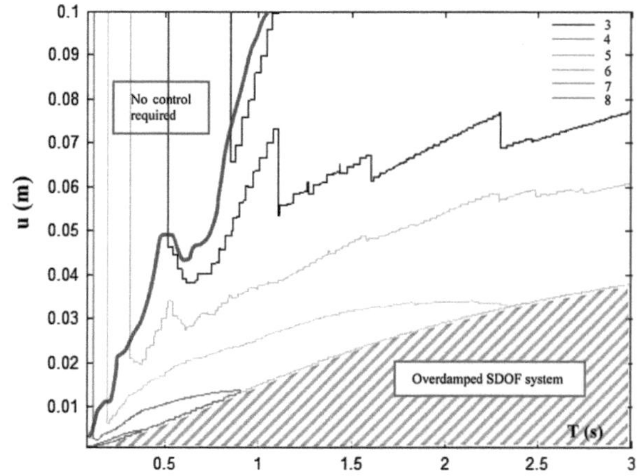

Fig. 6.103 Contour lines of total normalized cost of the optimal integrated structural system for different values of period and performance

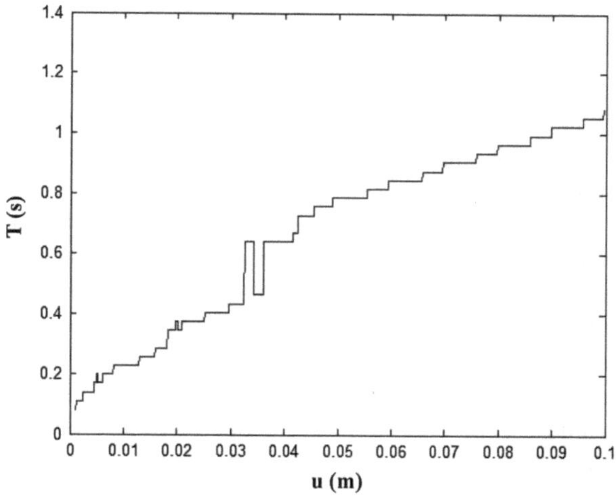

Fig. 6.104 Economically optimal period for different performance values

The periods, for which the normalized cost of the integrated structural system is minimum, are represented in Fig. 6.104 for different performance values.

From Fig. 6.104, it is clear that, since the relative ratio C_v/C_s of this case is high, like in the previous case, there is a regular trend to choose integrated structural systems having a less high period with less high viscoelastic damping resources resulting the most economically convenient.

With reference both to the normalized cost of the optimal unbraced structural system as well as the normalized cost of the optimal braced structural system evaluated for different values of the period T (i.e., undamped natural frequency ω) and seismic performance u, Figs. 6.65, 6.66, 6.67 and 6.68 remain unchanged.

In Fig. 6.105, the minimum values of the normalized cost of the integrated structural system $\overline{C}_{tot}$, the normalized ones of both unbraced $\overline{C}_s$ and braced $\overline{C}_{sb}$ structures relative to the optimal periods are represented.

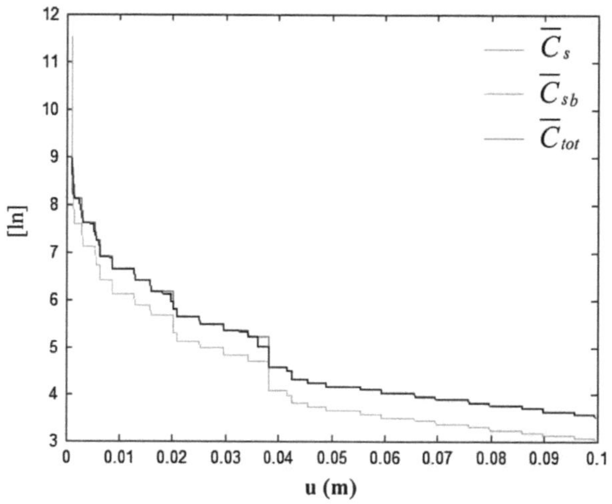

Fig. 6.105 Comparison between normalized costs related to the optimal design solutions considered for different performance values

It can be noted that the integrated system, constituted by the structural system and the viscoelastic dissipative bracing system, is never cheaper than the solution of the braced structural system and it is cheaper than the solution of the unbraced structural system for very few of the seismic design displacements considered.

The constraint on the maximum value of the overall damping ratio, also in this case, is respected.

With reference to the cases 1, 2, and 3, as shown in Figs. 6.69, 6.76, and 6.83, relative to the economic optimum of the integrated system obtained by considering the optimal vibration period, it is noted that the cost of the integrated system is always lower than the cost of the other two systems considered showing, thus, the convenience of viscoelastic dissipation in seismic design of structural systems. In such cases, as shown in Figs. 6.64, 6.75 and 6.82, there is a trend for different values of the seismic performance to choose integrated structural systems with a high period and with high viscoelastic damping resources resulting the most economically convenient. This trend is also justified by the fact that as the period T increases and, thus, the natural frequency ω decreases, the efficacy of the reduction factor of the static viscosity coefficient increases.

With reference to case 4, as shown in Fig. 6.90, the integrated design is the cheapest, as highlighted in the first three cases, only for some seismic design displacements considered. However, the integrated design is cheaper than the design solution of the structural system with the function of transferring both the vertical loads as well as seismic actions. In fact, from Fig. 6.89 it is deduced that, since the relative ratio C_v/C_s of this case is medium, the trend to choose integrated structural systems with a high period and with high viscoelastic damping resources for different values of the seismic performance is less marked.

With reference to the cases 5, 6, and 7, as shown in Figs. 6.97, 6.101 and 6.105, relative to the economic optimum of the integrated system obtained by considering the optimal vibration period, it is noted that the cost of the integrated system tends to be more and more comparable to that of the structural system with the result that the braced system is the economically optimal solution. In fact, from Figs. 6.96, 6.100 and 6.104, it should be noted that, since the relative ratio C_v/C_s of this case is high, there is a regular trend to choose integrated structural systems with more lateral stiffness and less viscoelastic damping resources for different target performance values.

Considering the results and Figs. 6.72, 6.79, 6.86 and 6.93, where a comparison between the costs related to the integrated design and the costs of the other two conventional design solutions is developed by also imposing the upper limit on the dimensionless overall viscous damping ratio, it is evident that in cases 1, 2, 3 and 4 imposing the upper limit on damping ratio necessarily leads to choose structures with more lateral stiffness (decrease of the economically optimal period) with a consequential increase in costs so that not for all seismic design displacements is the integrated design cheaper than the braced structure. In fact, it is possible to deduce from Figs. 6.70, 6.77, 6.84 and 6.91, that, unlike the respective analysis without constraint on the damping ratio, there is a more gradual and less marked trend to choose structural integrated systems with a high period and high viscoelastic damping resources resulting more cost-effective.

With reference to cases 5, 6 and 7 (Figs. 6.97, 6.101 and 6.105), the costs associated to integrated design turn out to be already related to periods of structural systems which correspond to damping ratios lower than 40 %.

From the above-mentioned critical analysis of cost comparisons for the three considered design solutions, it is possible to conclude that for relatively low ratios of cost, the integrated design with supplemental viscoelastic energy dissipation is the cost-effective optimal solution for any seismic performance. In other cases, the solution of braced system can be the most convenient.

Note, however, that these results are relative to the geometric configuration of the viscous damper-brace component with an inclination angle equal to 45°, which, as discussed in Sect. 6.3.1, is the configuration with the lowest dynamic efficiency. Referring to the other configurations, also for example the one in which the device is horizontal, it follows that the efficiency of the damper is double or more than double, as it can be seen in the equations illustrated in Fig. 5.5, and, therefore, the values of the static stiffness would be reduced approximately by 50 % with the consequence that the costs curves related to integrated systems would be always lower than the other two curves of the conventional design solutions considered.

6.6 The Effectiveness of the Integrated Design Methodology Evaluated on a SDOF System

A numerical example of an equivalent single-degree-of-freedom (SDOF) system is used to illustrate the feasibility and effectiveness of the integrated design procedure by using the design abacuses proposed and illustrated in Figs. 6.106, 6.107 and 6.108.

Consider, therefore, a regular structure, whose fundamental vibration period, evaluated, for example, or through simple empirical relations or with reference to an existing structural system, is equal to T = 0.78 s and the natural undamped frequency equal to $\omega = 7.97$ rad/s, with a mass m, and a seismic design performance $u = 3$ cm is requested.

In order to have an economically optimal integrated design and by assuming the following ratios of cost (case 2): $C_b/C_s = 0.5$ and $C_v/C_s = 3$, the total lateral stiffness of the structure with this period and mass can be evaluated:

$$\mathrm{k} = 63.59\ \mathrm{N/m} \times \mathrm{mass}$$

From Fig. 6.106 (point A′), it is obtained that the ratio of the economically optimal lateral stiffness of the structural system divided by the total lateral stiffness of the integrated structural system is equal to:

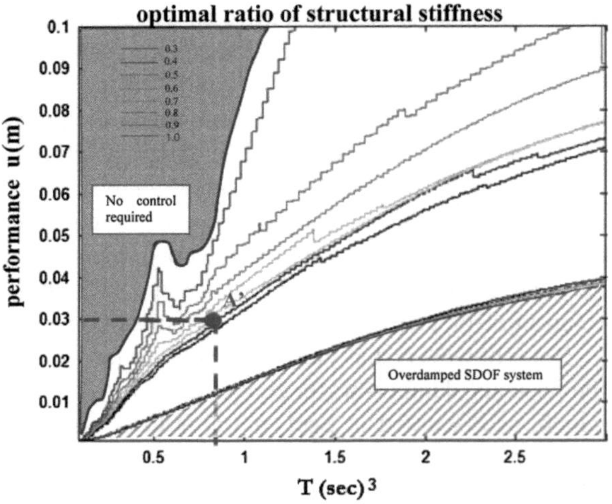

Fig. 6.106 Contour lines of optimal ratio k_s/k

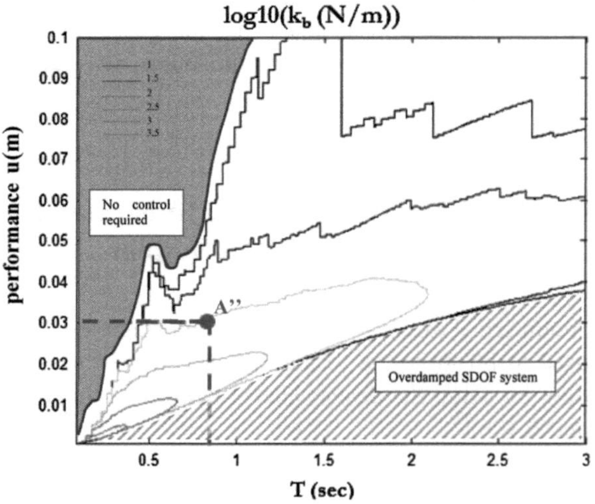

Fig. 6.107 Contour lines of optimal value of k_b related to optimal ratio k_s/k

$$k_s/k = 0.57$$

and it follows that the value of the lateral stiffness of the structural system is:

$$k_s = k_{s\ optimum} = 36.12\ \text{N/m} \times \text{mass}$$

By using Figs. 6.107 and 6.108 (points A″ and A‴), the above-mentioned economically optimal relationship of stiffness corresponds to the two economically optimal values of the static stiffness and static damping coefficient of the visco-elastic dissipative bracing system, with an inclined angle equal to $\vartheta = \pi/4$, which are respectively:

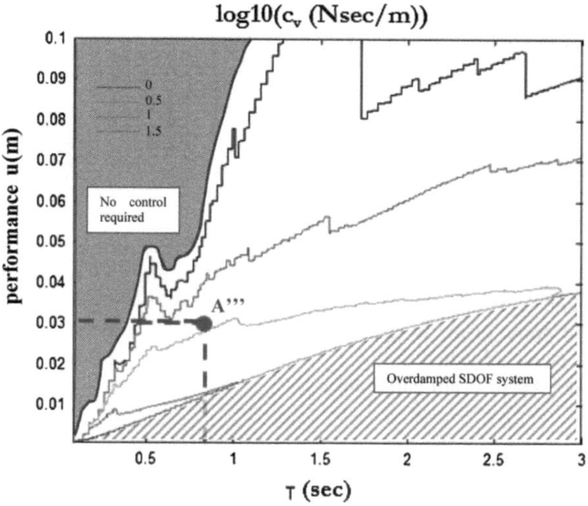

Fig. 6.108 Contour lines of optimal value of c_v related to optimal ratio k_s/k

$$k_b = k_{b\,\text{optimum}} = 50.61\ \text{N/m} \times \text{mass}$$

$$c_v = c_{v\,\text{optimum}} = 6.96\ \text{Ns/m} \times \text{mass}$$

It follows that the relaxation time is:

$$\tau = c_{v\,\text{optimum}}/k_{b\,\text{optimum}} = 0.14\ \text{s}$$

From Fig. 6.11 the value of the overall viscous damping ratio $\xi = 12\ \%$ ($\xi_s = 2\,\%$ and $\xi_d = 10\,\%$) is known. It is thus possible to estimate the fundamental damped frequency $\omega_D = 7.92$ rad/s (Eq. (5.4)) of the integrated structural system and, finally, the corresponding values of the dynamic damping coefficient and stiffness of the viscoelastic bracing-damper system, in accordance with the Eqs. (4.41) and (4.42), which are equal to:

$$c_{app} = c_{vb}(\omega_D) = c'(\omega_D) = c_v \frac{1}{1+\tau^2\omega_D^2} = 3.19\ \text{Ns/m} \times \text{mass}$$

$$k_{app}\cos^2\vartheta = k_{vb}(\omega_D)\cos^2\vartheta = k'(\omega_D)\cos^2\vartheta = \left(k_b \frac{\tau^2\omega_D^2}{1+\tau^2\omega_D^2}\right)\cos^2\vartheta$$
$$= 27.47\,\text{N/m} \times \text{mass}$$

In Fig. 6.109, the model of the integrated system is shown.

Finally, the response of the integrated structural system subjected to the set of the seven acceleration records considered has been analysed to assess whether the expected seismic design displacement is achieved in average. Figures 6.110, 6.111,

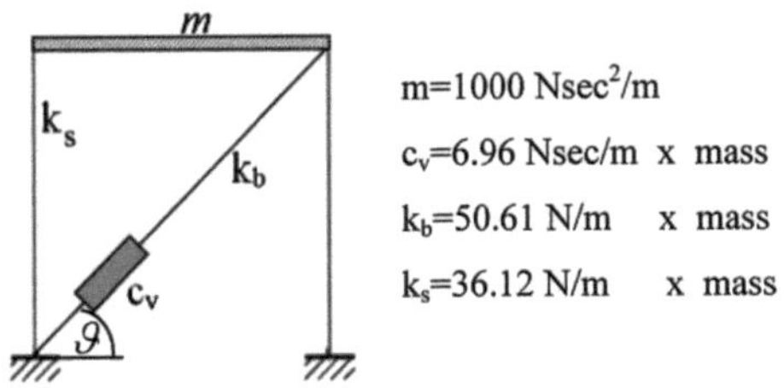

Fig. 6.109 Analytical model of the optimal integrated SDOF structure

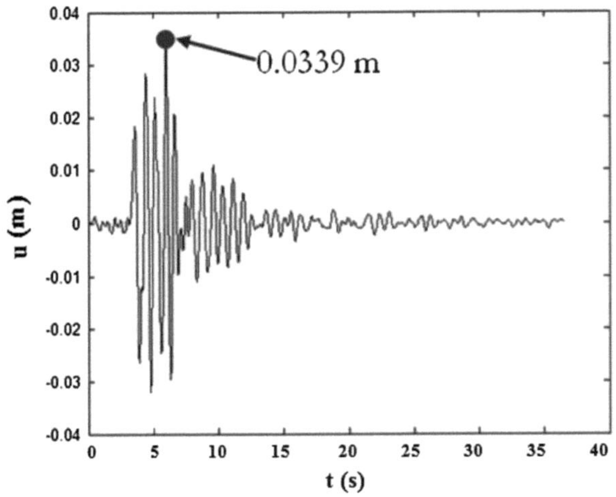

Fig. 6.110 The dynamic response of the integrated system subjected to the 000055xa-Friuli earthquake

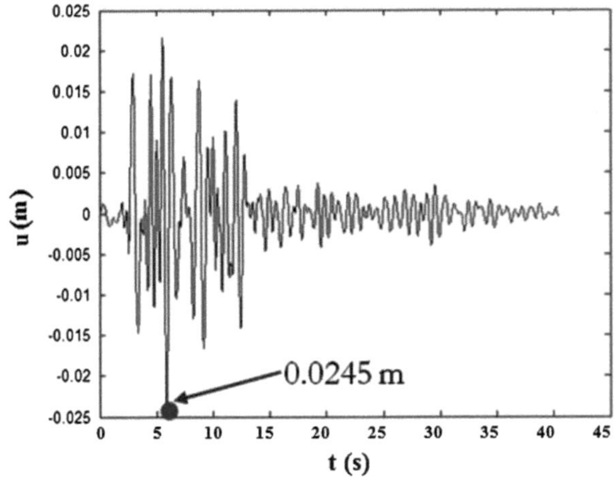

Fig. 6.111 The dynamic response of the integrated system subjected to the 000198xa-Montenegro earthquake

6.112, 6.113, 6.114, 6.115 and 6.116 show the time responses in terms of displacement to the seven accelerometric recordings.

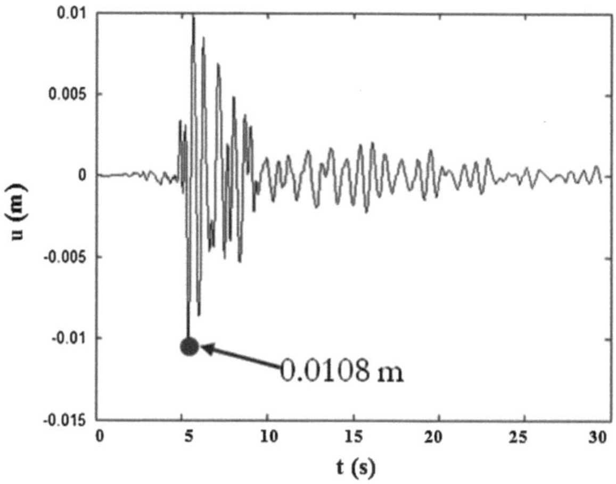

Fig. 6.112 The dynamic response of the integrated system subjected to the 000665xa-Umbria Marche earthquake

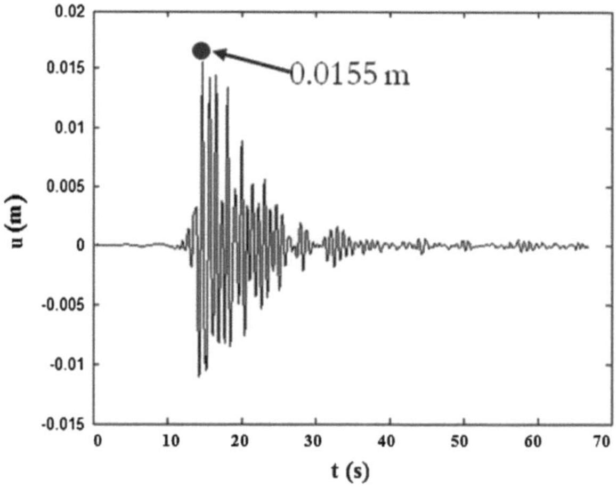

Fig. 6.113 The dynamic response of the integrated system subjected to the 004675ya-South Iceland earthquake

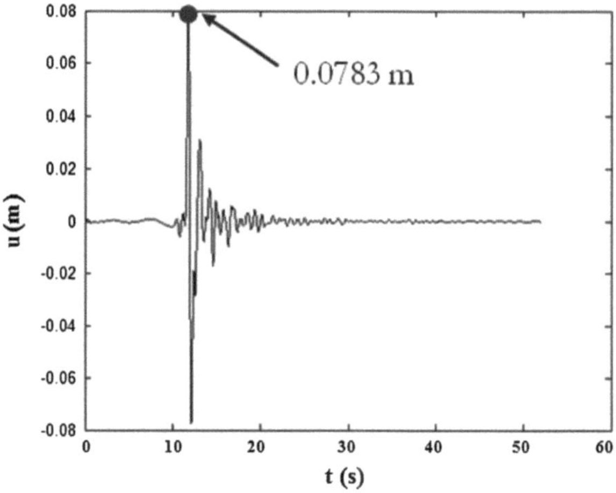

Fig. 6.114 The dynamic response of the integrated system subjected to the 006332ya-South Iceland (after shock) earthquake

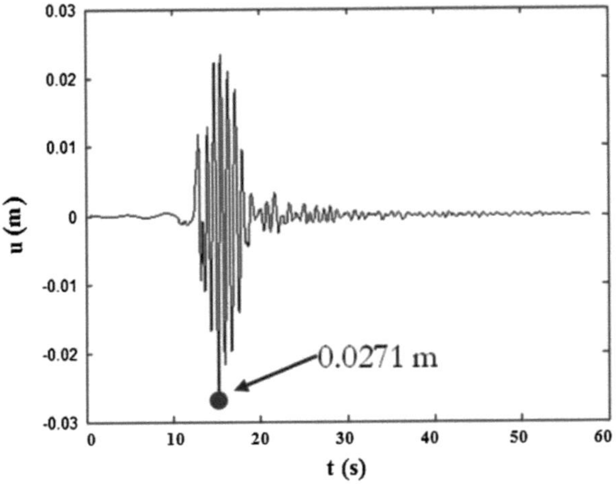

Fig. 6.115 The dynamic response of the integrated system subjected to the 006335xa-South Iceland (after shock) earthquake

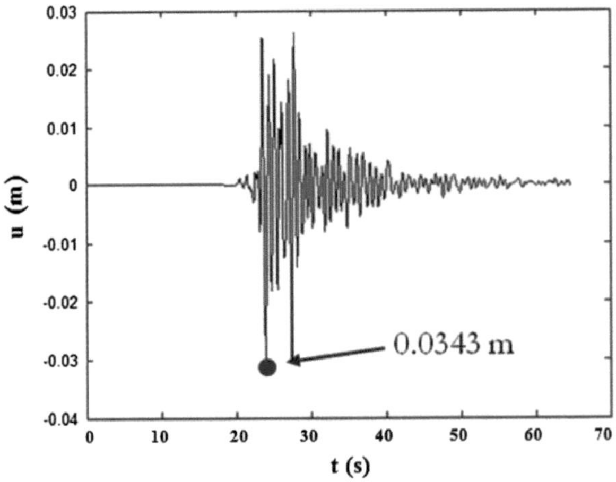

Fig. 6.116 The dynamic response of the integrated system subjected to the 007142ya-Bingol earthquake

Table 6.1 The dynamic response of the integrated system subjected to the seven accelerometric recordings selected

Earthquake	u (cm)
000055xa-Friuli	3.39
000198xa-Montenegro	2.45
000665xa-Umbria Marche	1.08
004675ya-South Iceland	1.55
006332ya-South Iceland (after shock)	7.83
006335xa-South Iceland (after shock)	2.71
007142ya-Bingol	3.43
u (average)	3.16

The average displacement is equal to u (average) = 3.16 cm. In Table 6.1, there is a summary of the results.

From Fig. 6.73 or 6.74, the normalized cost of the integrated structure considered is equal to $\bar{C}_{tot} = 107.63$ multiplied by the mass.

Subsequently, the response of a structure without supplemental damping (i.e., $\xi = 2\ \%$), which is able to achieve the expected performance, has been analysed. The period T has been chosen as high as possible in order to have the minimum value of the normalized cost: period T = 0.43 s or undamped natural frequency $\omega = 14.49$ rad/s. This structure is representative of both unbraced and braced structures.

The total lateral stiffness is:

$$k = 209.87\ \text{N/m} \times \text{mass}$$

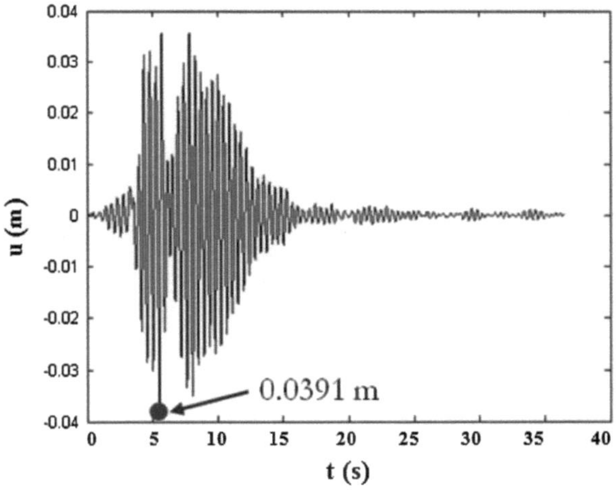

Fig. 6.117 The dynamic response of the structural system without damper subjected to the 000055xa-Friuli earthquake

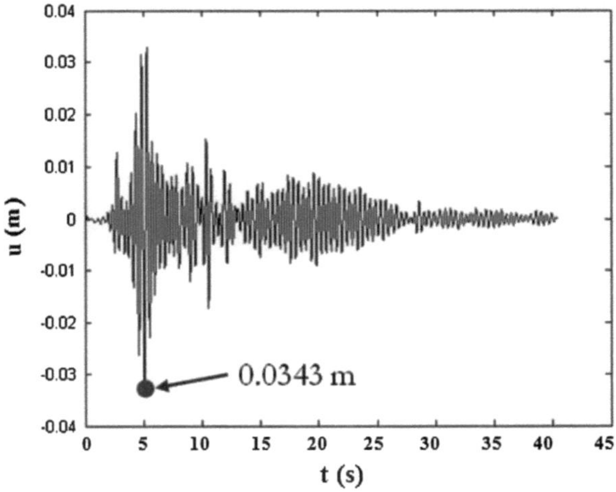

Fig. 6.118 The dynamic response of the structural system without damper subjected to the 000198xa-Montenegro earthquake

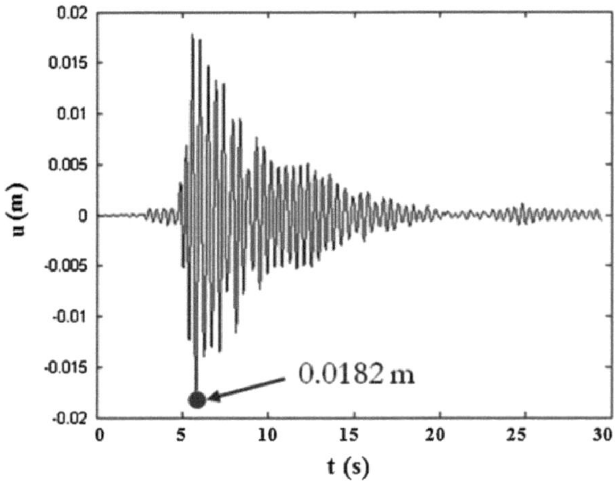

Fig. 6.119 The dynamic response of the structural system without damper subjected to the 000665xa-Umbria Marche earthquake

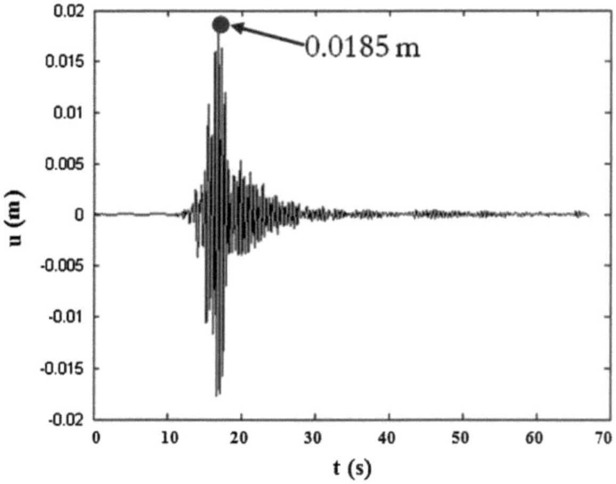

Fig. 6.120 The dynamic response of the structural system without damper subjected to the 004675ya-South Iceland earthquake

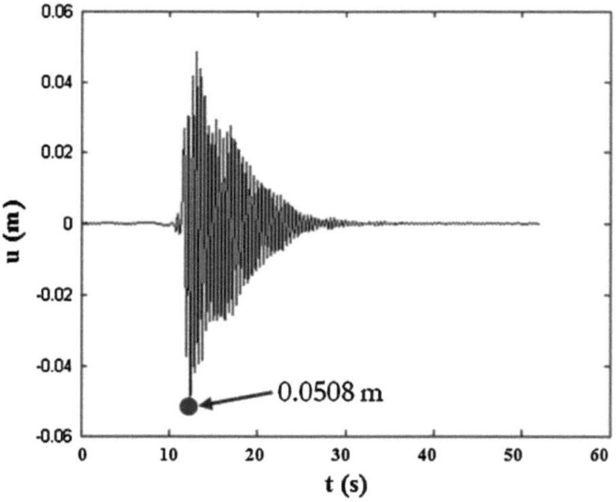

Fig. 6.121 The dynamic response of the structural system without damper subjected to the 006332ya-South Iceland (after shock) earthquake

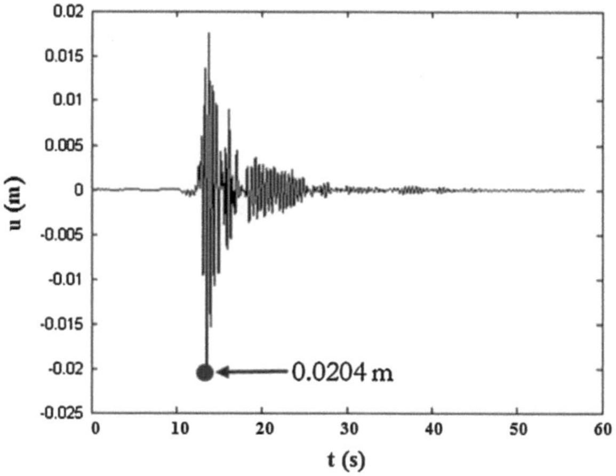

Fig. 6.122 The dynamic response of the structural system without damper subjected to the 006335xa-South Iceland (after shock) earthquake

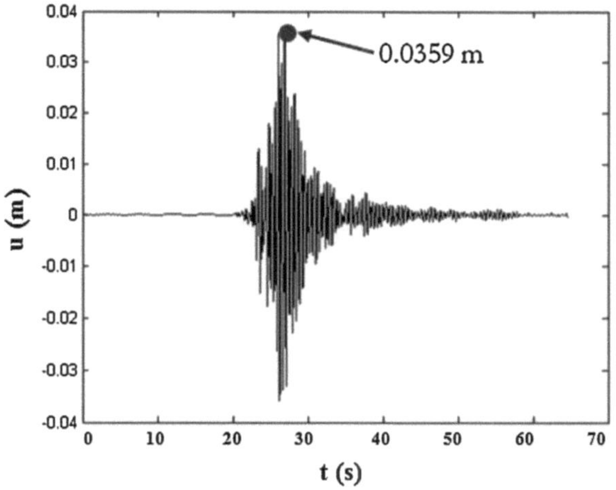

Fig. 6.123 The dynamic response of the structural system without damper subjected to the 007142ya-Bingol earthquake

Figures 6.117, 6.118, 6.119, 6.120, 6.121, 6.122 and 6.123 show the time responses in terms of displacement to the seven accelerometric records.

The average displacement is equal to u (average) = 3.10 cm. In Table 6.2 a summary of the results is reported.

From Fig. 6.76, the normalized cost of the unbraced structure considered is equal to $\bar{C}_s = 209.87$ multiplied by the mass; the normalized cost of the braced structure is equal to $\bar{C}_{sb} = 125.92$ multiplied by the mass.

Both values are higher than the normalized cost of the optimal integrated structural system with period T = 0.78 s demonstrating the cost-effectiveness of the solution found through the proposed integrated design methodology.

In these analyses, the period of the structure is considered fixed, i.e. it has not been treated as a design variable.

Table 6.2 The dynamic response of the structural system without damper subjected to the seven accelerometric recordings selected

Earthquake	u (cm)
000055xa-Friuli	3.91
000198xa-Montenegro	3.43
000665xa-Umbria Marche	1.82
004675ya-South Iceland	1.85
006332ya-South Iceland (after shock)	5.08
006335xa-South Iceland (after shock)	2.04
007142ya-Bingol	3.59
u (average)	3.10

By removing this hypothesis and including T between the design parameters, it is possible to find even cheaper solutions. In fact, with reference to the analyses carried out considering the constraint on the overall damping ratio of 40 %, the optimal period is equal to T = 1.05 s with $\xi = 26$ % and a corresponding normalized cost $\bar{C}_{tot} = 103.75$ relative to a unit mass, as seen in Fig. 6.79.

6.7 The Effectiveness of the Integrated Design Methodology Evaluated on a MDOF System with a Uniform Distribution of Stiffness

In this section, the design procedure developed on an equivalent SDOF integrated system will be extended to a proportionally damped MDOF framed integrated system on the basis of the hypotheses explained in Sect. 5.6 and by using the design abacuses proposed and illustrated in Sect. 6.4. This is the last step of the proposed integrated design methodology (Sect. 5.4, Fig. 5.5) in order to design the least expensive regular MDOF framed structure. In particular, this extension is illustrated through an application approach based on the results of the case study described in Sect. 6.6. It should be noted that such an extension is possible both in the case of designing new structures (i.e., the vibration period is not known a priori), as well as in the case of the existing structure with period known.

Consider, therefore, a three-storey regular structure (Fig. 6.124), whose fundamental vibration period, in analogy to the previous case, is equal to $T_1 = 0.78$ s or undamped natural frequency equal to $\omega_1 = 7.97$ rad/s, and having a mass "$m = 1000$ Ns^2/m" per level, and a seismic design performance, in terms of relative displacement between the top level and the ground, equal to u = 3 cm is requested. The relationship of the cost of the case 2, $C_b/C_s = 0.5$ and $C_v/C_s = 3$, an inclination angle of the viscoelastic dissipative bracing system, in each of the three structural levels, equal to $\vartheta = \pi/4$, are always considered.

Fig. 6.124 MDOF system equipped with linear viscous dampers

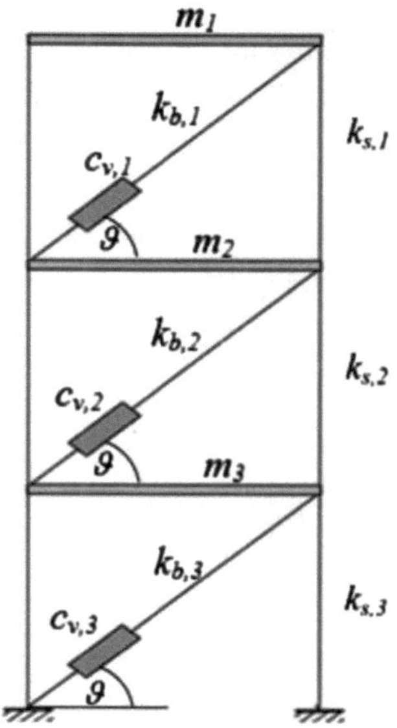

By writing the motion equations of the three levels in the time domain and defining k_1 the stiffness of the last level and k_3 the stiffness of the first level, as shown in Fig. 6.124, it follows that the displacement of each level u_i is the relative displacement between that level and the ground. The system of motion Eq. (6.6) is:

$$\begin{aligned} &m_1\,\ddot{u}_1(t) + k_1u_1(t) - k_1u_2(t) = f_1(t) \\ &m_2\,\ddot{u}_2(t) - k_1u_1(t) + k_1u_2(t) + k_2u_2(t) - k_2u_3(t) = f_2(t) \\ &m_3\,\ddot{u}_3(t) - k_2u_2(t) + k_2u_3(t) + k_3u_3(t) = f_3(t) \end{aligned} \tag{6.6}$$

that in matrix form becomes:

$$\begin{bmatrix} m_1 & 0 & 0 \\ 0 & m_2 & 0 \\ 0 & 0 & m_3 \end{bmatrix}\begin{bmatrix} \ddot{u}_1(t) \\ \ddot{u}_2(t) \\ \ddot{u}_3(t) \end{bmatrix} + \begin{bmatrix} k_1 & -k_1 & 0 \\ -k_1 & k_1+k_2 & -k_2 \\ 0 & -k_2 & k_2+k_3 \end{bmatrix}\begin{bmatrix} u_1(t) \\ u_2(t) \\ u_3(t) \end{bmatrix} = \begin{bmatrix} f_1(t) \\ f_2(t) \\ f_3(t) \end{bmatrix} \tag{6.7}$$

Since the mass and the lateral stiffness of each level are respectivily m and k, in order to have the frequency of the first mode equal to $\omega_1 = 7.97$ rad/s, the value of the stiffness k must be equal to 320800 N/m, as shown in Fig. 6.125 in which the trend of the stiffness of a generic floor is represented for different values of the square of the frequency of the first mode with reference to a structure with three levels and a mass $m = 1000$ Ns^2/m per level.

The mass matrix **M** is:

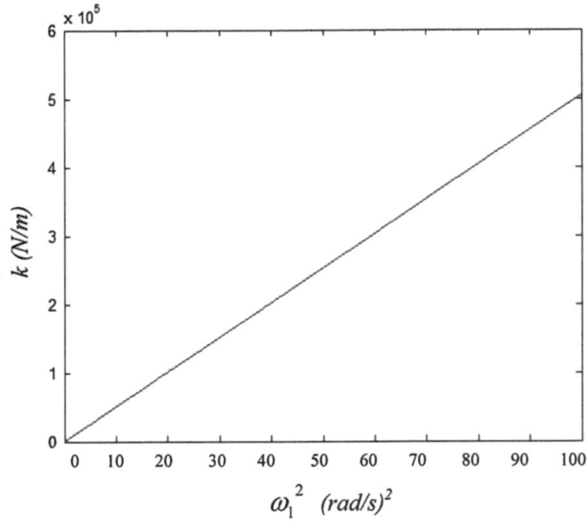

Fig. 6.125 The lateral stiffness k for different values of ω_1^2

$$\mathbf{M} = \begin{bmatrix} m_1 & 0 & 0 \\ 0 & m_2 & 0 \\ 0 & 0 & m_3 \end{bmatrix} = \begin{bmatrix} 1000 & 0 & 0 \\ 0 & 1000 & 0 \\ 0 & 0 & 1000 \end{bmatrix} \text{Ns}^2/\text{m} \tag{6.8}$$

Note that the increase of the masses produced by the viscoelastic dissipative bracing system is considered negligible.

Then the total lateral stiffness matrix **K** of the integrated system is:

$$\mathbf{K} = \begin{bmatrix} k_1 & -k_1 & 0 \\ -k_1 & k_1 + k_2 & -k_2 \\ 0 & -k_2 & k_2 + k_3 \end{bmatrix} = \begin{bmatrix} 320800 & -320800 & 0 \\ -320800 & 641600 & -320800 \\ 0 & -320800 & 641600 \end{bmatrix} \text{N/m} \tag{6.9}$$

By solving the homogeneous system:

$$\mathbf{Mu(t)} + \mathbf{Ku(t)} = \mathbf{0} \tag{6.10}$$

since the matrices are symmetric, for the Spectral Theorem, from the diagonalization of the following matrix:

$$\det|\mathbf{K} - \lambda\mathbf{M}| = \mathbf{0} \tag{6.11}$$

the eigenvectors $\boldsymbol{\varphi}_i$, which constitute the basis of the modal space, and the associated eigenvalues λ_i are obtained. It is possible to write the eigenvectors $\boldsymbol{\varphi}_i$ of the mode shapes in columns of a square matrix $\boldsymbol{\Phi}$.

It is, then, possible to normalize these eigenvectors by imposing the first component of each eigenvector (Eq. (6.12)) equal to 1 or normalize them with respect to $\boldsymbol{\Phi}^T\mathbf{M}\boldsymbol{\Phi} = \mathbf{I}$.

$$\mathbf{\Phi} = \begin{bmatrix} 1.00 & 1.00 & 1.00 \\ 0.80 & -0.56 & -2.25 \\ 0.44 & -1.25 & 1.80 \end{bmatrix} \tag{6.12}$$

The eigenvectors represent the shapes of the corresponding modes. The eigenvalues are the squares of the frequencies of the relative modes: $\lambda_i = \omega_i^2$. The diagonal matrix of the eigenvalues $\mathbf{\Omega^2}$ (Eq. (6.13)) can, similarly, be obtained.

$$\Omega^2 = \begin{bmatrix} \omega_1^2 & 0 & 0 \\ 0 & \omega_2^2 & 0 \\ 0 & 0 & \omega_3^2 \end{bmatrix} = \begin{bmatrix} 63.54 & 0 & 0 \\ 0 & 498.83 & 0 \\ 0 & 0 & 1041.6 \end{bmatrix} (\text{rad/s})^2 \tag{6.13}$$

Making the modal transformation of coordinates from normal coordinates (i.e., displacements at the nodes) into modal coordinates:

$$\mathbf{u} = \mathbf{\Phi y} \tag{6.14}$$

it is possible to calculate the generalized mass matrix:

$$\mathbf{\Phi^T M \Phi} = \begin{bmatrix} m_1^* & 0 & 0 \\ 0 & m_2^* & 0 \\ 0 & 0 & m_3^* \end{bmatrix} = \begin{bmatrix} 1841.2 & 0 & 0 \\ 0 & 2862.9 & 0 \\ 0 & 0 & 9295.9 \end{bmatrix} \text{Ns}^2/\text{m} \tag{6.15}$$

the generalized stiffness matrix:

$$\mathbf{\Phi^T K \Phi} = \begin{bmatrix} k_1^* & 0 & 0 \\ 0 & k_2^* & 0 \\ 0 & 0 & k_3^* \end{bmatrix} = 10^6 \begin{bmatrix} 0.12 & 0.00 & -0.00 \\ 0.00 & 1.43 & 0.00 \\ -0.00 & 0.00 & 9.68 \end{bmatrix} \text{N/m} \tag{6.16}$$

the generalized force vector:

$$\mathbf{\Phi^T F}(t) = \begin{bmatrix} f_1^*(t) \\ f_2^*(t) \\ f_3^*(t) \end{bmatrix} = -\mathbf{\Phi^T}(\mathbf{Mr})\ddot{u}_g(t) \tag{6.17}$$

where

$$\mathbf{r} = \begin{bmatrix} 1 \\ 1 \\ 1 \end{bmatrix} \tag{6.18}$$

The participation factor of the i-th mode is defined as:

$$g_i = \frac{\boldsymbol{\varphi}_i^T(\mathbf{Mr})}{\boldsymbol{\varphi}_i^T \mathbf{M} \boldsymbol{\varphi}_i} = \frac{\boldsymbol{\varphi} \boldsymbol{i}^{\mathbf{T}}(\mathbf{Mr})}{m_i^*} \tag{6.19}$$

Then, the vector of participation factors turns out to be:

$$\mathbf{g} = \begin{bmatrix} 1.22 \\ -0.28 \\ 0.06 \end{bmatrix} \tag{6.20}$$

As regards the period $T_1 = 0.78$ s or $\omega_1 = 7.97$ rad/s and the performance equal to "u/g_1", from Figs. 6.11, 6.28, 6.29 and 6.30, the following values apply:

- $\xi_1 = 22\,\%$
- $k_s/k = 0.21$
- $c_{v\,\text{optimum}} = 12.95\,\text{Ns/m} \times \text{mass}$
- $k_{b\,\text{optimum}} = 99.11\,\text{N/m} \times \text{mass}$
- $\tau = 0.13$ s
- $\omega_{D,1} = \omega_1\sqrt{1-\xi_1^2} = 7.78$ rad/s

On basis of the economically optimal relationship of the lateral stiffness of the structural system divided by the total lateral stiffness of the integrated structural system k_s/k, that is 0.21, it follows that the lateral stiffness matrix of the structural system $\mathbf{K_s}$ is:

$$\mathbf{K}_s = 0.21\mathbf{K} \tag{6.21}$$

the dynamic stiffness matrix $\mathbf{K'_b}$ of the viscoelastic dissipative bracing system, indicated $\bar{\mathbf{K}}$ in Chap. 3, is the complementary and, therefore, is the following:

$$\mathbf{K'_b}\cos^2\vartheta = \mathbf{K} - \mathbf{K_s} \tag{6.22}$$

Knowing the relaxation time τ and the damped frequency of the first mode, in accordance with the Eq. (4.41), the static stiffness matrix of the viscoelastic dissipative bracing system $\mathbf{K_b}$ can be obtained by using the following equation:

$$\mathbf{K_b} = \mathbf{K'_b}\left(\frac{1+\tau^2\omega_{1,D}^2}{\tau^2\omega_{1,D}^2}\right) \tag{6.23}$$

With reference to the damping, it can be, in general, expressed by the definition of a damping matrix proportional to the mass matrix and/or stiffness one (Rayleigh proportional damping) or by directly specifying damping values in the uncoupled motion equations written in the modal space as proposed by Wilson (discrete modal damping) (Clough and Penzien 1993; Wilson 2000). By following these approaches, the system is called proportionally damped system.

With reference to the inherent damping matrix of the structure, a Rayleigh damping matrix has been adopted proportional both to the mass and stiffness matrices in order to obtain a damping ratio equal to 2 % on the first and second modes:

$$\mathbf{C_s} = \alpha\mathbf{M} + \beta\mathbf{K} = \begin{bmatrix} 651.9 & -417 & 0 \\ -417 & 1069 & -417 \\ 0 & -417 & 1069 \end{bmatrix} \text{Ns/m} \quad (6.24)$$

where

$$\begin{aligned} \beta &= \xi\frac{2}{(\omega_1 + \omega_2)} = 0.0013 \\ \alpha &= \beta(\omega_1\omega_2) = 0.235 \end{aligned} \quad (6.25)$$

With reference to the dynamic supplemental damping matrix $\mathbf{C'_v}$ of the viscoelastic dissipative bracing system, a Rayleigh damping matrix has been adopted so that the corresponding static supplemental damping matrix is proportional only to the static stiffness matrix of the viscoelastic dissipative system via the proportionality factor: relaxation time τ (Eq. (6.26)). The proportionality only with the stiffness matrix has been chosen both in order to have a dynamic behavior of the dissipative system in analogy to that studied on the substitute SDOF system as well as to have a profile of the velocities similar to the one of the displacements.

$$\mathbf{C_v} = \tau\mathbf{K_b} \quad (6.26)$$

It is possible to obtain the dynamic supplemental damping matrix $\mathbf{C'_v}$ of the viscoelastic dissipative bracing system, indicated as $\bar{\mathbf{C}}$ in Chap. 3, in accordance with Eq. (4.42), by using the following equation:

$$\begin{aligned} \mathbf{C'_v}\cos^2\vartheta &= \left(C_v\frac{1}{(1+\tau^2\omega_{1,D}^2)}\right)\cos^2\vartheta \\ &= \begin{bmatrix} c'_1 & -c'_1 & 0 \\ -c'_1 & c'_1 + c'_2 & -c'_2 \\ 0 & -c'_2 & c'_2 + c'_3 \end{bmatrix}\cos^2\vartheta = \begin{bmatrix} 16083 & -16083 & 0 \\ -16083 & 32166 & -16083 \\ 0 & -16083 & 32166 \end{bmatrix} \text{Ns/m} \end{aligned} \quad (6.27)$$

In this way, the overall damping matrix $\mathbf{C} = \mathbf{C_s} + \mathbf{C'_v}\cos^2\vartheta$ is diagonalizable with the same basis of eigenvectors, since the criterion of Caughey is also verified:

$$\mathbf{CM}^{-1}\mathbf{K} = \mathbf{KM}^{-1}\mathbf{C}$$

Making the modal transformation of coordinates from normal coordinates (i.e., displacements at the nodes) into modal coordinates (Eq. (6.14)), the generalized damping matrix can be calculated:

$$\mathbf{\Phi}^{\mathrm{T}}\mathbf{C}\mathbf{\Phi} = \begin{bmatrix} c_1^* & 0 & 0 \\ 0 & c_2^* & 0 \\ 0 & 0 & c_3^* \end{bmatrix} = 10^5\begin{bmatrix} 0.06 & 0.00 & -0.00 \\ 0.00 & 0.72 & 0.00 \\ -0.00 & 0.00 & 4.85 \end{bmatrix} \text{Ns/m} \quad (6.28)$$

It follows that the modal space equations are decoupled because all the generalized matrices are diagonal and, therefore, the system of equations becomes:

$$
\begin{aligned}
m_1^* \ddot{y_1}(t) + c_1^* \dot{y}_1(t) + k_1^* y_1(t) &= f_1^*(t) \\
m_2^* \ddot{y_2}(t) + c_1^* \dot{y}_2(t) + k_2^* y_2(t) &= f_2^*(t) \\
m_3^* \ddot{y_3}(t) + c_1^* \dot{y}_3(t) + k_3^* y_3(t) &= f_3^*(t)
\end{aligned}
\tag{6.29}
$$

in matrix form:

$$
\boldsymbol{\Phi}^{\mathbf{T}}\mathbf{M}\boldsymbol{\Phi}\,\ddot{\mathbf{y}} + \boldsymbol{\Phi}^{\mathbf{T}}\mathbf{C}\boldsymbol{\Phi}\dot{\mathbf{y}} + \boldsymbol{\Phi}^{\mathbf{T}}\mathbf{K}\boldsymbol{\Phi}\,\mathbf{y} = \boldsymbol{\Phi}^{\mathbf{T}}\mathbf{F}(t) \tag{6.30}
$$

Dividing the Eq. (6.29) by the i-th generalized mass, it is obtained:

$$
\begin{aligned}
\ddot{y_1}(t) + 2\xi_1\omega_1\dot{y}_1(t) + \omega_1^2 y_1(t) &= -g_1\,\ddot{u}_g(t) \\
\ddot{y_2}(t) + 2\xi_2\omega_2\dot{y}_2(t) + \omega_2^2 y_2(t) &= -g_2\,\ddot{u}_g(t) \\
\ddot{y_3}(t) + 2\xi_3\omega_3\dot{y}_3(t) + \omega_3^2 y_3(t) &= -g_3\,\ddot{u}_g(t)
\end{aligned}
\tag{6.31}
$$

The damping ratios are:

$$
\boldsymbol{\xi} = \begin{bmatrix} 22\,\% \\ 58\,\% \\ 81\,\% \end{bmatrix} \tag{6.32}
$$

The effective modal masses can be evaluated as follows:

$$
\widehat{m}_i = \boldsymbol{\varphi}_i^{\mathbf{T}}(\mathbf{Mr})\frac{\boldsymbol{\varphi}_i^{\mathbf{T}}(\mathbf{Mr})}{\boldsymbol{\varphi}_i^{\mathbf{T}}\mathbf{M}\boldsymbol{\varphi}_i} = \boldsymbol{\varphi}_i^{\mathbf{T}}(\mathbf{Mr})\frac{\boldsymbol{\varphi}_i^{\mathbf{T}}(\mathbf{Mr})}{\boldsymbol{\varphi}_i^{\mathbf{T}}\mathbf{M}\boldsymbol{\varphi}_i}\frac{\boldsymbol{\varphi}_i^{\mathbf{T}}(\mathbf{M}\boldsymbol{\varphi}_i)}{\boldsymbol{\varphi}_i^{\mathbf{T}}(\mathbf{M}\boldsymbol{\varphi}_i)} = \boldsymbol{\varphi}_i^{\mathbf{T}}\mathbf{M}\boldsymbol{\varphi}_i g_i^2 \tag{6.33}
$$

The effective modal mass matrix is equal to:

$$
\widehat{\mathbf{M}} = \begin{bmatrix} \widehat{m}_1 & 0 & 0 \\ 0 & \widehat{m}_2 & 0 \\ 0 & 0 & \widehat{m}_3 \end{bmatrix} = \begin{bmatrix} 2742.2 & 0 & 0 \\ 0 & 224.6 & 0 \\ 0 & 0 & 33.1 \end{bmatrix} \mathrm{Ns^2/m} \tag{6.34}
$$

The sum of the effective modal masses of all the modes must give the sum of the masses of all the floors of the building that is 3000 $\mathrm{Ns^2/m}$.

The percentage of each effective modal mass turns out to be:

$$
\frac{\widehat{\mathbf{M}}}{3m} = \begin{bmatrix} \frac{\widehat{m}_1}{3m} & 0 & 0 \\ 0 & \frac{\widehat{m}_2}{3m} & 0 \\ 0 & 0 & \frac{\widehat{m}_3}{3m} \end{bmatrix} = \begin{bmatrix} 91.41\,\% & -0.00\,\% & 0.00\,\% \\ -0.00\,\% & 7.49\,\% & 0.00\,\% \\ 0\% & 0.00\,\% & 1.10\,\% \end{bmatrix} \tag{6.35}
$$

The trace of the above-mentioned matrix is equal to 100 %. Note that only the effective mass related to the first mode exceeds 90 %, the minimum value that must be achieved by summing the effective modal masses of the modes considered (BSSC 2004).

In order to carry out the dynamic analyses, the whole procedure has also been developed, as shown in Chap. 3, in the time domain both in terms of relative motions of the mass and in state space form.

With reference to the 000055xa-Friuli accelerometric registration, in Figs. 6.126, 6.127 and 6.128, by having examined the contribution only of the first vibration mode of the integrated structural system, the viscoelastic and viscous responses, respectively, of the linear viscous damper and viscous damper-brace component for each level are shown and compared to the global viscoelastic response of the corresponding structural level by considering the horizontal projection of all the forces.

By applying the modal strain energy method on MDOF systems, illustrated in Sect. 3.2.1, and using the Eqs. (3.55)–(3.57), it has been possible to estimate the energy dissipated by linear viscous dampers and the elastic energy of the system relative to the first mode shape of the system, which result being:

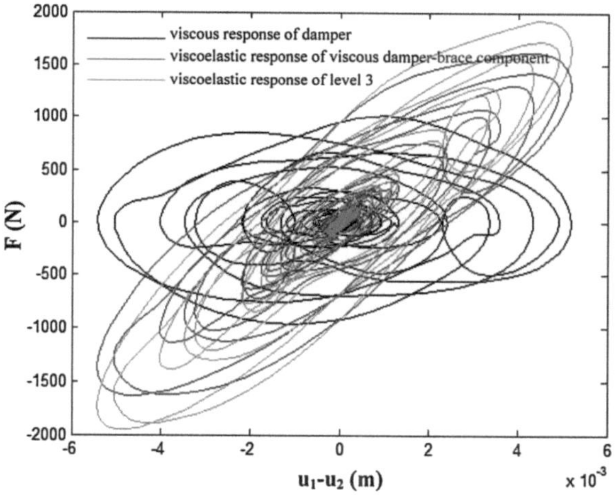

Fig. 6.126 Hysteresis loops at level 3, evaluated by considering the contribution only of the first vibration mode, related to the 000055xa-Friuli earthquake

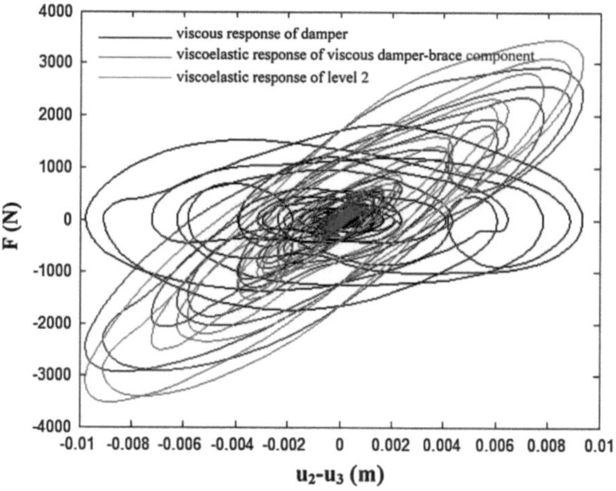

Fig. 6.127 Hysteresis loops at level 2, evaluated by considering the contribution only of the first vibration mode, related to the 000055xa-Friuli earthquake

Fig. 6.128 Hysteresis loops at level 1, evaluated by considering the contribution only of the first vibration mode, related to the 000055xa-Friuli earthquake

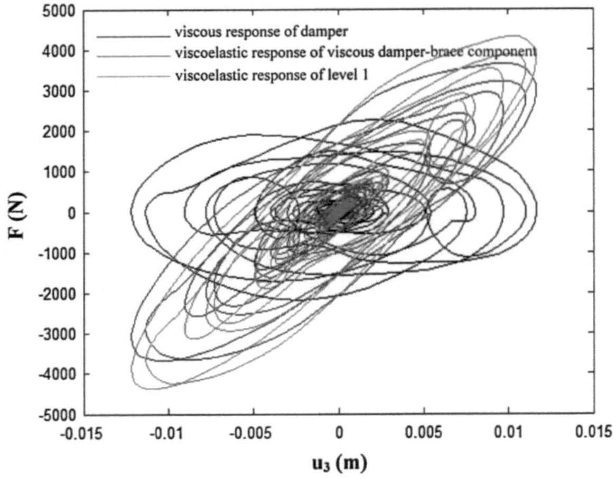

$$\sum_j W_j = \frac{2\pi^2}{T_1}\sum_j c'\left(\omega_{D,1}\right)_j \phi_{d,j}^2 \cos^2 \vartheta_j = 1.47 \cdot 10^5\,\text{Nm}$$

$$W_k = \Phi_1^T \mathbf{K}\Phi_1 = \Phi_1^T \omega_1^2 \mathbf{M}\Phi_1 = \sum_i \omega_1^2 m_i \phi_i^2 = \frac{4\pi^2}{T_1^2}\sum_i m_i \phi_i^2 = 1.17 \cdot 10^5\,\text{Nm}$$

it has been therefore, possible, by using the Eqs. (3.54) and (3.58), to evaluate the damping ratio of the first mode, which results being:

$$\xi_{d,1} = \frac{\frac{2\pi^2}{T_1}\sum\limits_j c'\left(\omega_{D,1}\right)_j \phi_{r,j}^2 \cos^2 \vartheta_j}{2\pi \frac{4\pi^2}{T_1^2}\sum\limits_i m_i \phi_i^2} = \frac{T_1 \sum\limits_j c'_j \phi_{r,j}^2 \cos^2 \vartheta_j}{4\pi \sum\limits_i m_i \phi_i^2} = 0.20$$

In accordance with Eq. (3.53), the overall damping ratio is:

$$\xi_1 = \xi_{s,1} + \xi_{d,1} = 22\,\%$$

It is emphasized that the application of the modal strain energy method, used for evaluation of the damping ratio of the first mode, has expressly taken into account the dynamic behavior of the viscoelastic system by employing the dynamic values of the viscosity coefficients and stiffnesses of the each viscous damper-brace component of each structural level assessed and considering the damped frequency of the first mode of the integrated structural system.

With reference to the same 000055xa-Friuli accelerometric recording, in Figs. 6.129, 6.130 and 6.131, by having examined the contribution of all the vibration modes of the integrated structural system, the viscoelastic and viscous responses, respectively, of the linear viscous damper and viscous damper-brace component for each level are shown and compared to the global viscoelastic

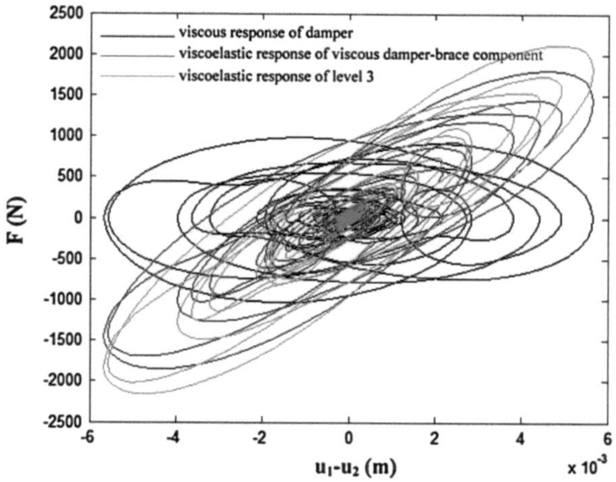

Fig. 6.129 Hysteresis loops at level 3, evaluated by considering the contributions of all the vibration modes, related to the 000055xa-Friuli earthquake

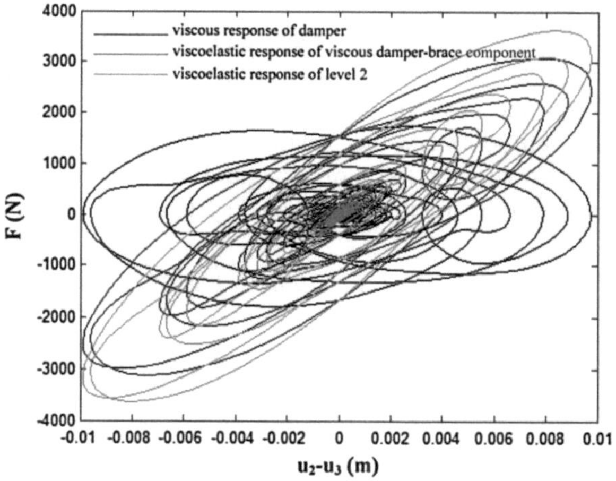

Fig. 6.130 Hysteresis loops at level 2, evaluated by considering the contributions of all the vibration modes, related to the 000055xa-Friuli earthquake

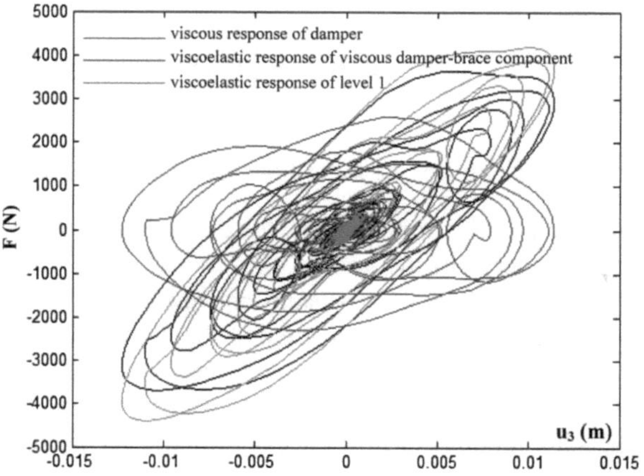

Fig. 6.131 Hysteresis loops at level 1, evaluated by considering the contributions of all the vibration modes, related to the 000055xa-Friuli earthquake

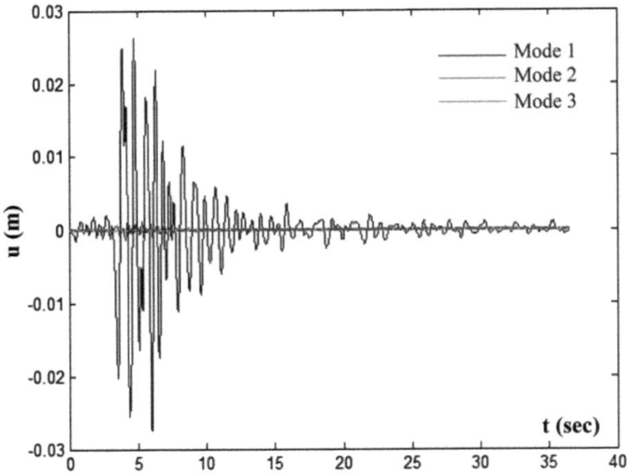

Fig. 6.132 Contributions of all vibration modes related to 000055xa-Friuli earthquake

response of the corresponding structural level by considering the horizontal projection of all the forces.

With reference to the same 000055xa-Friuli accelerometric recording, Fig. 6.132 shows the contributions of the three modes.

On the basis of the Eqs. (1.6)–(1.11), (1.13), (6.6)–(6.9), (6.21)–(6.27), it is possible to extend the "relative" energy balance (Bertero and Uang 1992; Fajfar and Krawinkler 2005) of a single-degree-of-freedom system to the multi-degrees-of-freedom structural integrated system, as follows:

$$\int_t \dot{\mathbf{u}}^{\mathrm{T}}\mathbf{M}\,\ddot{\mathbf{u}}dt + \int_t \dot{\mathbf{u}}^{\mathrm{T}}\mathbf{C}_{\mathrm{s}}\dot{\mathbf{u}}dt + \int_t \dot{\mathbf{u}}^{\mathrm{T}}\left(\mathbf{C}'_{\mathrm{v}}\cos^2\vartheta\right)\dot{\mathbf{u}}dt + \int_t \dot{\mathbf{u}}^{\mathrm{T}}\mathbf{K}_{\mathrm{s}}\mathbf{u}dt + \int_t \dot{\mathbf{u}}^{\mathrm{T}}\left(\mathbf{K}'_{\mathrm{b}}\cos^2\vartheta\right)\mathbf{u}dt = -\int_t \dot{\mathbf{u}}^{\mathrm{T}}\mathbf{M}\,\ddot{\mathbf{u}}_{\mathrm{g}}dt \tag{6.36}$$

$$\frac{1}{2}\dot{\mathbf{u}}^{\mathrm{T}}\mathbf{M}\underline{\mathbf{u}} + \int_t \dot{\mathbf{u}}^{\mathrm{T}}\mathbf{C}_{\mathrm{s}}\dot{\mathbf{u}}dt + \int_t \dot{\mathbf{u}}^{\mathrm{T}}\left(\mathbf{C}'_{\mathrm{v}}\cos^2\vartheta\right)\dot{\mathbf{u}}dt + \frac{1}{2}\mathbf{u}^{\mathrm{T}}\mathbf{K}_{\mathrm{s}}\mathbf{u} + \frac{1}{2}u^{T}\left(\mathbf{K}'_{\mathrm{b}}\cos^2\vartheta\right)\mathbf{u} = -\int_t \dot{\mathbf{u}}^{\mathrm{T}}\mathbf{M}\,\ddot{\mathbf{u}}_{g}dt \tag{6.37}$$

$$E_K + E_{D,\,structure} + E_{D,\,devices} + E_{E,\,structure} + E_{E,\,braces} = E_I \tag{6.38}$$

in which, in the first member, the different rates, evaluated in relative terms, of the kinetic energy of the system, the viscous energy both of the structural system that the dissipative bracing system and the elastic energy of both systems are combined, the whose sum must equal the cumulative energy input.

In Fig. 6.133, the above-mentioned accumulated energy rates of the integrated structural system are represented for different values of the time, with reference to the same 000055xa-Friuli accelerometric recording.

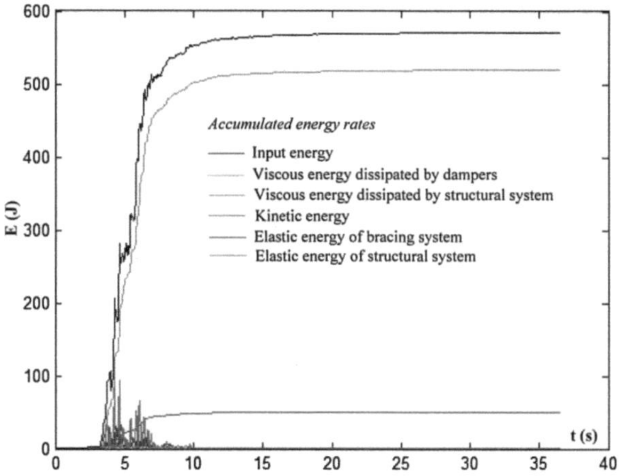

Fig. 6.133 Accumulated energy rates of the MDOF integrated structural system related to the 000055xa-Friuli earthquake

Then, by considering the seven accelerometric recordings, the responses in terms of relative displacements between each level of the MDOF system and the ground are illustrated in Figs. 6.134, 6.135, 6.136, 6.137, 6.138, 6.139 and 6.140.

In Table 6.3, the maximum values of the response of the MDOF integrated structural system subjected to the seven accelerometric recordings considered and the corresponding average value are shown.

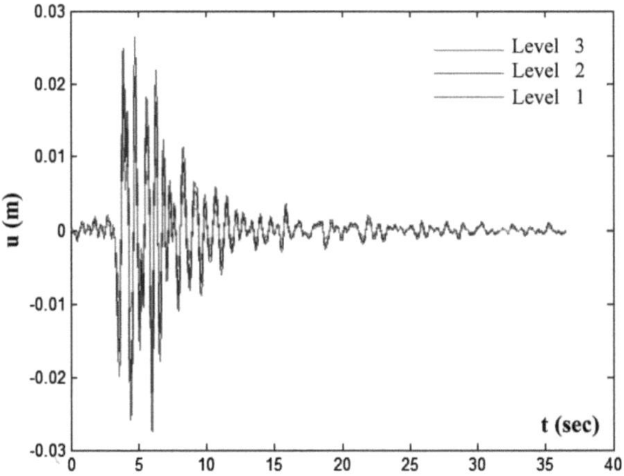

Fig. 6.134 The dynamic response of the MDOF integrated system subjected to the 000055xa-Friuli earthquake

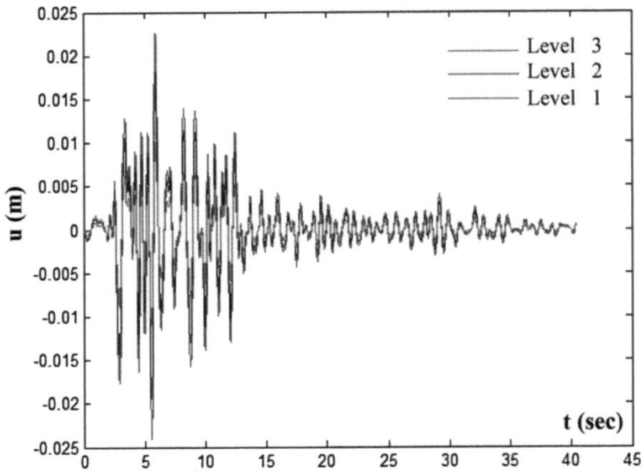

Fig. 6.135 The dynamic response of the MDOF integrated system subjected to the 000198xa-Montenegro earthquake

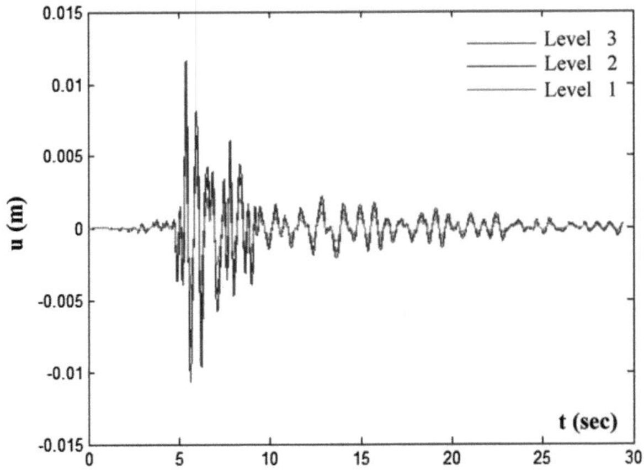

Fig. 6.136 The dynamic response of the MDOF integrated system subjected to the 000665xa-Umbria Marche earthquake

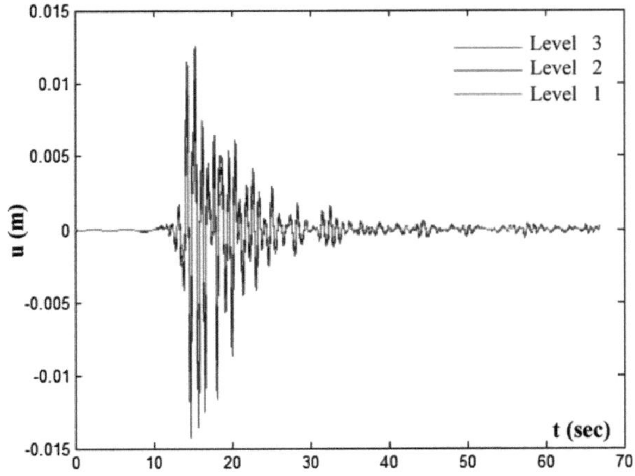

Fig. 6.137 The dynamic response of the MDOF integrated system subjected to the 004675ya-South Iceland earthquake

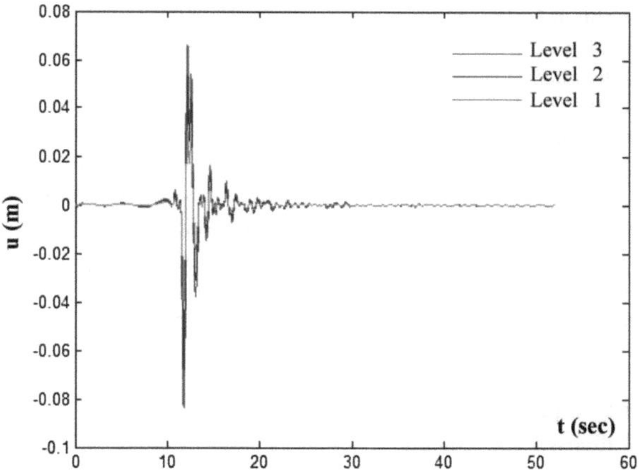

Fig. 6.138 The dynamic response of the MDOF integrated system subjected to the 006332ya-South Iceland (after shock) earthquake

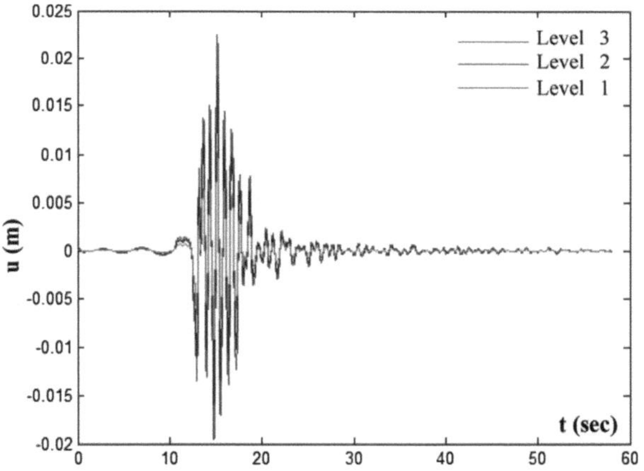

Fig. 6.139 The dynamic response of the MDOF integrated system subjected to the 006335xa-South Iceland (after shock) earthquake

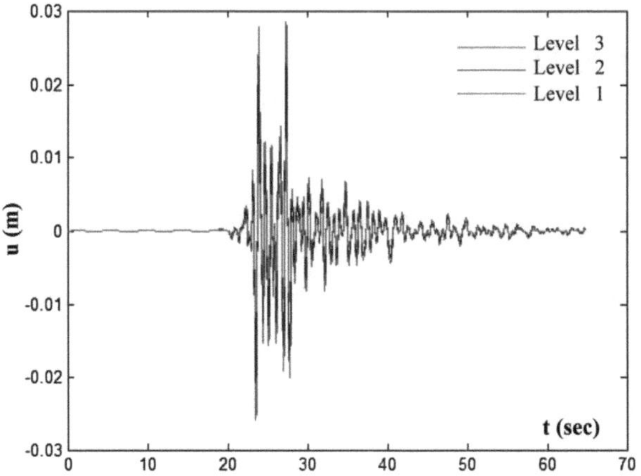

Fig. 6.140 The dynamic response of the MDOF integrated system subjected to the 007142ya-Bingol earthquake

Table 6.3 The dynamic response of the MDOF integrated system subjected to the seven accelerometric recordings selected

Earthquake	u (cm)
000055xa-Friuli	2.74
000198xa-Montenegro	2.41
000665xa-Umbria Marche	1.17
004675ya-South Iceland	1.42
006332ya-South Iceland (after shock)	8.33
006335xa-South Iceland (after shock)	2.24
007142ya-Bingol	2.86
u (average)	3.02

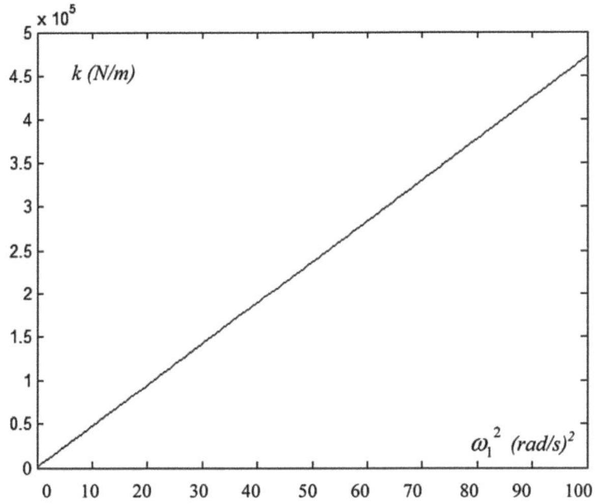

Fig. 6.141 The lateral stiffness k for different values of ω_1^2

The average displacement is equal to 3.02 cm confirming the effectiveness of the integrated design methodology.

6.8 The Effectiveness of the Integrated Design Methodology Evaluated on a MDOF System with a Non Uniform Distribution of Stiffness

In this section, a MDOF framed integrated structure similar to the one (Fig. 6.124) described in the previous paragraph is illustrated with the only difference being 20 % variations of stiffness between two consecutive floors. This stiffness variation does not exceed the limits indicated by NTC08 (NTC 2008), and, so the structure can still be defined regular.

The system of motion equations is similar to Eq. (6.6).

The mass of each level is $m_1 = m_2 = m_3 = m$. By defining $k_2 = k$ the lateral stiffness of the second level, the lateral stiffness of the uppest level and first level one will be, respectivily, $k_3 = k/1.2$ and $k_1 = 1.2\ k$. In order to have the frequency of the first mode equal to $\omega_1 = 7.97$ rad/s, the value of the stiffness k must be equal to 300000 N/m, as shown in Fig. 6.141 in which the trend of the stiffness of the second floor is represented for different values of the square of the frequency of the first mode with reference to a structure with three levels and a mass m = 1000 Ns^2/m per level and 20 % variations of stiffness between two consecutive floors.

The mass matrix **M** turns out to be completely equal to Eq. (6.8).

Then, the total lateral stiffness matrix **K** of the integrated system is:

$$\mathbf{K} = \begin{bmatrix} k_1 & -k_1 & 0 \\ -k_1 & k_1 + k_2 & -k_2 \\ 0 & -k_2 & k_2 + k_3 \end{bmatrix} = 10^4 \begin{bmatrix} 25 & -25 & 0 \\ -25 & 55 & -30 \\ 0 & -30 & 66 \end{bmatrix} \mathrm{N/m} \qquad (6.39)$$

The matrix $\mathbf{\Phi}$ of the eigenvectors normalized by imposing the first component of each eigenvector equal to 1 results being (Eq. (6.40)):

$$\mathbf{\Phi} = \begin{bmatrix} 1.00 & 1.00 & 1.00 \\ 0.74 & -0.79 & -2.80 \\ 0.37 & -1.10 & 2.90 \end{bmatrix} \qquad (6.40)$$

The eigenvectors represent the shapes of the corresponding modes. The eigenvalues are the squares of the frequencies of the relative modes: $\lambda_i = \omega_i^2$. The diagonal matrix of the eigenvalues $\mathbf{\Omega^2}$ (Eq. (6.41)) can, similarly, be obtained.

$$\mathbf{\Omega^2} = \begin{bmatrix} \omega_1^2 & 0 & 0 \\ 0 & \omega_2^2 & 0 \\ 0 & 0 & \omega_3^2 \end{bmatrix} = \begin{bmatrix} 63.66 & 0 & 0 \\ 0 & 446.50 & 0 \\ 0 & 0 & 949.83 \end{bmatrix} (\mathrm{rad/s})^2 \qquad (6.41)$$

Making the modal transformation of coordinates from normal coordinates (i.e., displacements at the nodes) into modal coordinates:

$$\mathbf{u} = \mathbf{\Phi y} \qquad (6.42)$$

it is possible to calculate the generalized mass matrix:

$$\mathbf{\Phi}^{\mathrm{T}}\mathbf{M\Phi} = \begin{bmatrix} m_1^* & 0 & 0 \\ 0 & m_2^* & 0 \\ 0 & 0 & m_3^* \end{bmatrix} = \begin{bmatrix} 1696 & 0 & 0 \\ 0 & 2838 & 0 \\ 0 & 0 & 17232 \end{bmatrix} \mathrm{Ns^2/m} \qquad (6.43)$$

the generalized stiffness matrix:

$$\mathbf{\Phi}^{\mathrm{T}}\mathbf{K\Phi} = \begin{bmatrix} k_1^* & 0 & 0 \\ 0 & k_2^* & 0 \\ 0 & 0 & k_3^* \end{bmatrix} = 10^7 \begin{bmatrix} 0.01 & 0.00 & 0.00 \\ 0.00 & 0.13 & -0.00 \\ 0.00 & -0.00 & 1.64 \end{bmatrix} \mathrm{N/m} \qquad (6.44)$$

the generalized force vector:

$$\mathbf{\Phi}^{\mathrm{T}}\mathbf{F}(t) = \begin{bmatrix} f_1^*(t) \\ f_2^*(t) \\ f_3^*(t) \end{bmatrix} = -\mathbf{\Phi}^{\mathrm{T}}(\mathbf{Mr})\ddot{u}_{\mathrm{g}}(t) \tag{6.45}$$

where

$$\mathbf{r} = \begin{bmatrix} 1 \\ 1 \\ 1 \end{bmatrix} \tag{6.46}$$

The participation factor of the i-th mode is defined as:

$$g_i = \frac{\boldsymbol{\varphi}_i^{\mathrm{T}}(\mathbf{Mr})}{\boldsymbol{\varphi}_i^{\mathrm{T}}\mathbf{M}\boldsymbol{\varphi}_i} = \frac{\boldsymbol{\varphi} i^{\mathrm{T}}(\mathbf{Mr})}{m_i^*} \tag{6.47}$$

Then, the vector of participation factors turns out to be:

$$\mathbf{g} = \begin{bmatrix} 1.25 \\ -0.31 \\ 0.06 \end{bmatrix} \tag{6.48}$$

This variation of the stiffness matrix will change the participation factor of the first mode g_1 as it can be seen from Eq. (6.48). Similarly to the previous case, with reference to the period $T_1 = 0.78$ s or $\omega_1 = 7.97$ rad/s and the seismic performance equal to u/g_1, from Figs. 6.11, 6.28, 6.29 and 6.30, the following values apply:

- $\xi_1 = 23\,\%$;
- $k_s/k = 0.21$;
- $c_{v\,\mathrm{optimum}} = 12.99\,\mathrm{Ns/m} \times \mathrm{mass}$;
- $k_{b\,\mathrm{optimum}} = 103.86\,\mathrm{N/m} \times \mathrm{mass}$;
- $\tau = 0.12$ s;
- $\omega_{D,1} = \omega_1\sqrt{1-\xi_1^2} = 7.76\ \mathrm{rad/s}$.

Similarly, the economically optimal relationship of the lateral stiffness of the structural system divided by the total lateral stiffness of the integrated structural system k_s/k, is 0.21 and it follows that the lateral stiffness matrix of the structural system $\mathbf{K_s}$ is:

$$\mathbf{K}_s = 0.21\mathbf{K} \tag{6.49}$$

the dynamic stiffness matrix $\mathbf{K'_b}$ of the viscoelastic dissipative bracing system is the complementary and, therefore, is the following:

$$\mathbf{K'}_{\mathrm{b}}\cos^2\vartheta = \mathbf{K} - \mathbf{K}_{\mathrm{s}} \tag{6.50}$$

As in the previous case, knowing the relaxation time τ and damped frequency of the first mode, in accordance with the Eq. (4.41), the static stiffness matrix of the viscoelastic dissipative bracing system $\mathbf{K_b}$ can be obtained by using the following equation:

$$\mathbf{K}_b = \mathbf{K'}_b\left(\frac{1+\tau^2\omega_{1,D}^2}{\tau^2\omega_{1,D}^2}\right) \tag{6.51}$$

As in the previous case, with reference to the inherent damping matrix of the structure, a Rayleigh damping matrix has been adopted proportional both to the mass and stiffness matrices in order to obtain a damping ratio equal to 2 % on the first and second modes:

$$\mathbf{C}_s = \alpha\mathbf{M} + \beta\mathbf{K} = \begin{bmatrix} 557.2 & -343.5 & 0 \\ -343.5 & 987.4 & -412.2 \\ 0 & -412.2 & 1138.6 \end{bmatrix} \text{Ns/m} \tag{6.52}$$

where

$$\begin{aligned} \beta &= \xi\frac{2}{(\omega_1+\omega_2)} = 0.0014 \\ \alpha &= \beta(\omega_1\omega_2) = 0.2317 \end{aligned} \tag{6.53}$$

With reference to the dynamic supplemental damping matrix $\mathbf{C'_v}$ of the viscoelastic dissipative bracing system, a Rayleigh damping matrix has been adopted so that the corresponding static supplemental damping matrix is proportional only to the static stiffness matrix of the viscoelastic dissipative bracing system via the proportionality factor: relaxation time τ.

$$\mathbf{C}_v = \tau\mathbf{K}_b \tag{6.54}$$

It is possible to obtain the dynamic supplemental damping matrix $\mathbf{C'_v}$ of the viscoelastic dissipative bracing system, in accordance with the Eq. (4.42), by using the following equation:

$$\begin{aligned} \mathbf{C}'_v\cos^2\vartheta &= \left(\mathbf{C}_v\frac{1}{(1+\tau^2\omega_{1,D}^2)}\right)\cos^2\vartheta \\ &= \begin{bmatrix} c'_1 & -c'_1 & 0 \\ -c'_1 & c'_1+c'_2 & -c'_2 \\ 0 & -c'_2 & c'_2+c'_3 \end{bmatrix}\cos^2\vartheta = \begin{bmatrix} 13157 & -13157 & 0 \\ -13157 & 28946 & -15789 \\ 0 & -15789 & 34735 \end{bmatrix}\text{Ns/m} \end{aligned} \tag{6.55}$$

In this way, the overall damping matrix $\mathbf{C} = \mathbf{C_s} + \mathbf{C'_v}\cos^2\vartheta$ is diagonalizable with the same basis of eigenvectors, since the criterion of Caughey is also verified:

$$\mathbf{CM^{-1}K} = \mathbf{KM^{-1}C}$$

Making the modal transformation of coordinates from normal coordinates (i.e., displacements at the nodes) into modal coordinates (Eq. (6.42)), the generalized damping matrix can be calculated:

$$\mathbf{\Phi}^{\mathrm{T}}\mathbf{C}\mathbf{\Phi} = \begin{bmatrix} c_1^* & 0 & 0 \\ 0 & c_2^* & 0 \\ 0 & 0 & c_3^* \end{bmatrix} = 10^5 \begin{bmatrix} 0.06 & 0.00 & 0.00 \\ 0.00 & 0.67 & -0.00 \\ 0.00 & -0.00 & 8.61 \end{bmatrix} \mathrm{Ns/m} \tag{6.56}$$

It follows that the modal space equations are decoupled because all the generalized matrices are diagonal and, therefore, the system of equations becomes:

$$\begin{aligned} m_1^* \ddot{y}_1(t) + c_1^* \dot{y}_1(t) + k_1^* y_1(t) &= f_1^*(t) \\ m_2^* \ddot{y}_2(t) + c_1^* \dot{y}_2(t) + k_2^* y_2(t) &= f_2^*(t) \\ m_3^* \ddot{y}_3(t) + c_1^* \dot{y}_3(t) + k_3^* y_3(t) &= f_3^*(t) \end{aligned} \tag{6.57}$$

in matrix form:

$$\mathbf{\Phi}^{\mathrm{T}}\mathbf{M}\mathbf{\Phi}\,\ddot{\mathbf{y}} + \mathbf{\Phi}^{\mathrm{T}}\mathbf{C}\mathbf{\Phi}\dot{\mathbf{y}} + \mathbf{\Phi}^{\mathrm{T}}\mathbf{K}\mathbf{\Phi}\,\mathbf{y} = \mathbf{\Phi}^{\mathrm{T}}\mathbf{F}(t) \tag{6.58}$$

Dividing the Eq. (6.57) by the i-th generalized mass, it is obtained:

$$\begin{aligned} \ddot{y_1}(t) + 2\xi_1\omega_1\dot{y}_1(t) + \omega_1^2 y_1(t) &= -g_1\,\ddot{u}_g(t) \\ \ddot{y_2}(t) + 2\xi_2\omega_2\dot{y}_2(t) + \omega_2^2 y_2(t) &= -g_2\,\ddot{u}_g(t) \\ \ddot{y_3}(t) + 2\xi_3\omega_3\dot{y}_3(t) + \omega_3^2 y_3(t) &= -g_3\,\ddot{u}_g(t) \end{aligned} \tag{6.59}$$

The damping ratios are:

$$\xi = \begin{bmatrix} 23\,\% \\ 58\,\% \\ 81\,\% \end{bmatrix} \tag{6.60}$$

The effective modal masses can be evaluated as follows:

$$\widehat{m}_i = \boldsymbol{\varphi}_i^{\mathrm{T}}(\mathbf{Mr})\frac{\boldsymbol{\varphi}_i^{\mathrm{T}}(\mathbf{Mr})}{\boldsymbol{\varphi}_i^{\mathrm{T}}\mathbf{M}\boldsymbol{\varphi}_i} = \boldsymbol{\varphi}_i^{\mathrm{T}}(\mathbf{Mr})\frac{\boldsymbol{\varphi}_i^{\mathrm{T}}(\mathbf{Mr})}{\boldsymbol{\varphi}_i^{\mathrm{T}}\mathbf{M}\boldsymbol{\varphi}_i}\frac{\boldsymbol{\varphi}_i^{\mathrm{T}}(\mathbf{M}\boldsymbol{\varphi}_i)^{\mathrm{T}}}{\boldsymbol{\varphi}\quad i}(\mathbf{M}\boldsymbol{\varphi}_i) = \boldsymbol{\varphi}_i^{\mathrm{T}}\mathbf{M}\boldsymbol{\varphi}_i g_i^2 \tag{6.61}$$

The effective modal mass matrix is equal to:

$$\widehat{\mathbf{M}} = \begin{bmatrix} \widehat{m}_1 & 0 & 0 \\ 0 & \widehat{m}_2 & 0 \\ 0 & 0 & \widehat{m}_3 \end{bmatrix} = \begin{bmatrix} 2650.6 & 0 & 0 \\ 0 & 279.5 & 0 \\ 0 & 0 & 70.0 \end{bmatrix} \mathrm{Ns^2/m} \tag{6.62}$$

The sum of the effective modal masses of all the modes must give the sum of the masses of all the floors of the building that is 3000 $\mathrm{Ns^2/m}$.

The percentage of each effective modal mass turns out to be:

$$\frac{\widehat{\mathbf{M}}}{3m} = \begin{bmatrix} \frac{\widehat{m}_1}{3m} & 0 & 0 \\ 0 & \frac{\widehat{m}_2}{3m} & 0 \\ 0 & 0 & \frac{\widehat{m}_3}{3m} \end{bmatrix} = \begin{bmatrix} 88.35\,\% & -0.00\,\% & 0.00\,\% \\ -0.00\,\% & 9.32\,\% & 0.00\,\% \\ 0\,\% & 0.00\,\% & 2.33\,\% \end{bmatrix} \tag{6.63}$$

the trace of the above-mentioned matrix is equal to 100 %.

The whole procedure has also been developed, as shown in Chap. 3, in the time domain both in terms of relative motions of the mass and in state space form.

As in the previous case, with reference to the 000055xa-Friuli accelerometric registration, in Figs. 6.142, 6.143 and 6.144, by having examined the contribution only of the first vibration mode of the integrated structural system, the viscoelastic and viscous responses, respectively, of the linear viscous damper and viscous damper-brace component for each level are shown and compared to the global viscoelastic response of the corresponding structural level by considering the horizontal projection of all the forces.

By applying the modal strain energy method on MDOF systems, as in the previous case, it has been possible to estimate the energy dissipated by linear viscous dampers and the elastic energy of the system relative to the first mode shape of the system, which result being:

$$\sum_j W_j = \frac{2\pi^2}{T_1} \sum_j c'\left(\omega_{D,1}\right)_j \phi_{d,j}^2 \cos^2 \vartheta_j = 1.42 \cdot 10^5\,\text{Nm}$$

$$W_k = \Phi_1^T \mathbf{K} \Phi_1 = \Phi_1^T \omega_1^2 \mathbf{M} \Phi_1 = \sum_i \omega_1^2 m_i \phi_i^2 = \frac{4\pi^2}{T_1^2} \sum_i m_i \phi_i^2 = 1.08 \cdot 10^5\,\text{Nm}$$

it has been, therefore, possible to evaluate the damping ratio of the first mode, which results being:

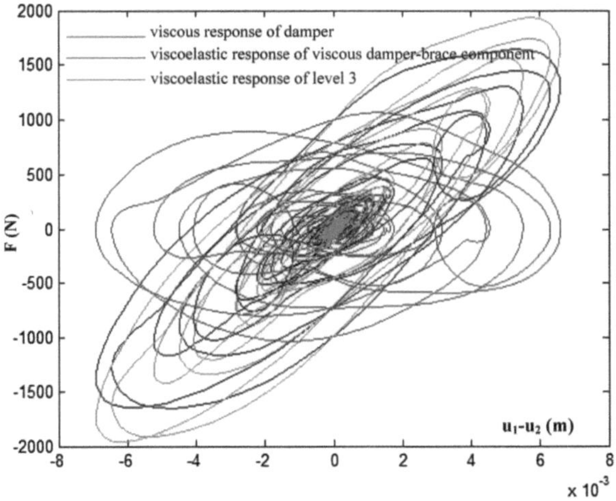

Fig. 6.142 Hysteresis loops at level 3, evaluated by considering the contribution only of the first vibration mode, related to the 000055xa-Friuli earthquake

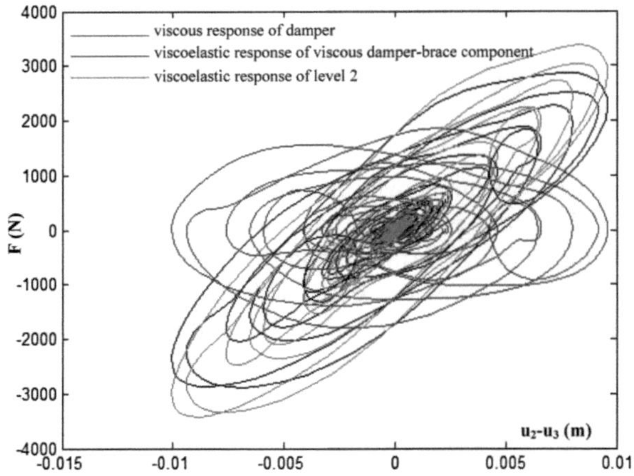

Fig. 6.143 Hysteresis loops at level 2, evaluated by considering the contribution only of the first vibration mode, related to the 000055xa-Friuli earthquake

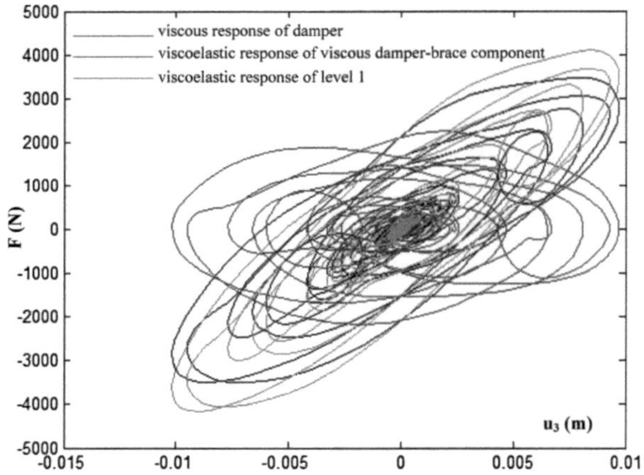

Fig. 6.144 Hysteresis loops at level 1, evaluated by considering the contribution only of the first vibration mode, related to the 000055xa-Friuli earthquake

$$\xi_{d,1} = \frac{\frac{2\pi^2}{T_1}\sum_j c'\left(\omega_{D,1}\right)_j \phi_{r,j}^2 \cos^2\vartheta_j}{2\pi\frac{4\pi^2}{T_1^2}\sum_i m_i\phi_i^2} = \frac{T_1\sum_j c'_j\phi_{r,j}^2\cos^2\vartheta_j}{4\pi\sum_i m_i\phi_i^2} = 0.21$$

In accordance with Eq. (3.53), the overall damping ratio is:

$$\xi_1 = \xi_{s,1} + \xi_{d,1} = 23\,\%$$

With reference to the same 000055xa-Friuli accelerometric recording, in Figs. 6.145, 6.146 and 6.147, by having examined the contribution of all the vibration modes of the integrated structural system, the viscoelastic and viscous responses, respectively, of the linear viscous damper and viscous damper-brace component for each level are shown and compared to the global viscoelastic response of the corresponding structural level by considering the horizontal projection of all the forces.

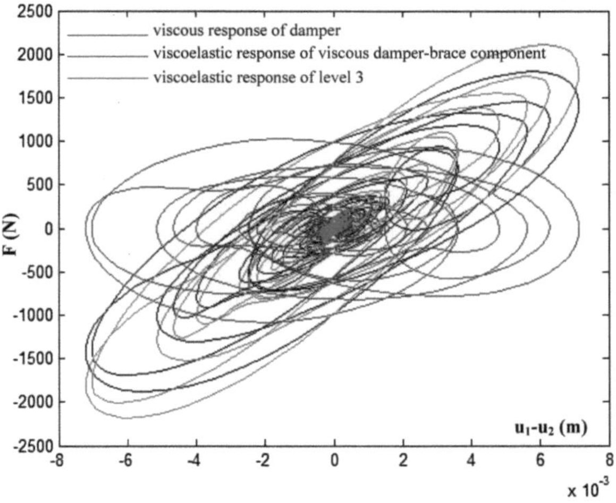

Fig. 6.145 Hysteresis loops at level 3, evaluated by considering the contributions of all the vibration modes, related to the 000055xa-Friuli earthquake

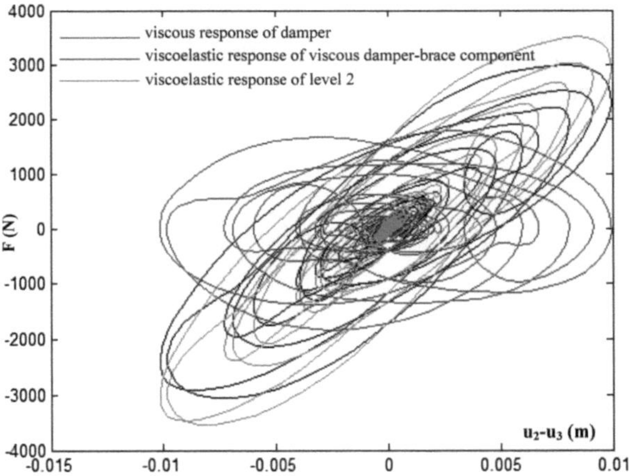

Fig. 6.146 Hysteresis loops at level 2, evaluated by considering the contributions of all the vibration modes, related to the 000055xa-Friuli earthquake

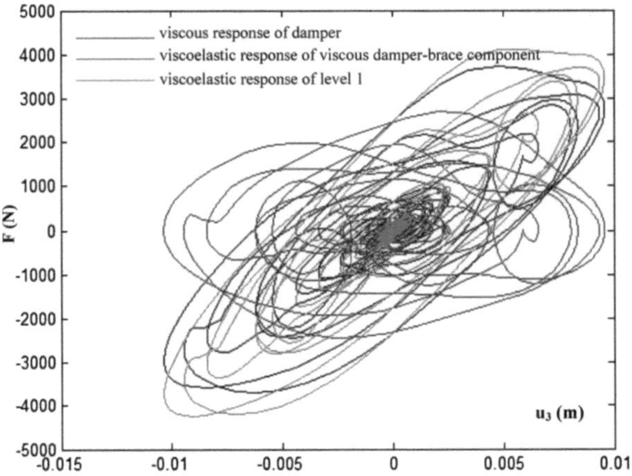

Fig. 6.147 Hysteresis loops at level 1, evaluated by considering the contributions of all the vibration modes, related to the 000055xa-Friuli earthquake

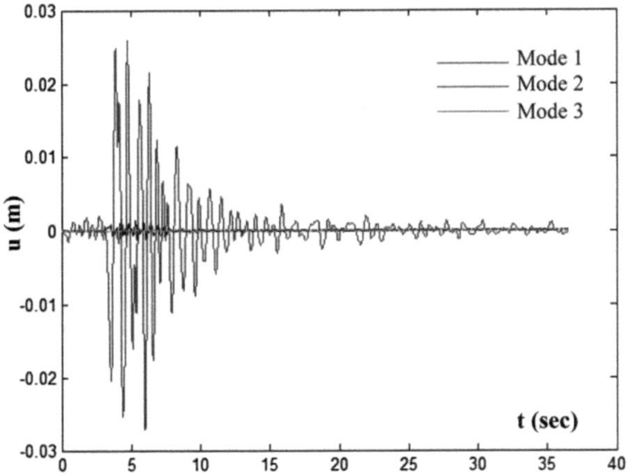

Fig. 6.148 Contributions of all vibration modes related to 000055xa-Friuli earthquake

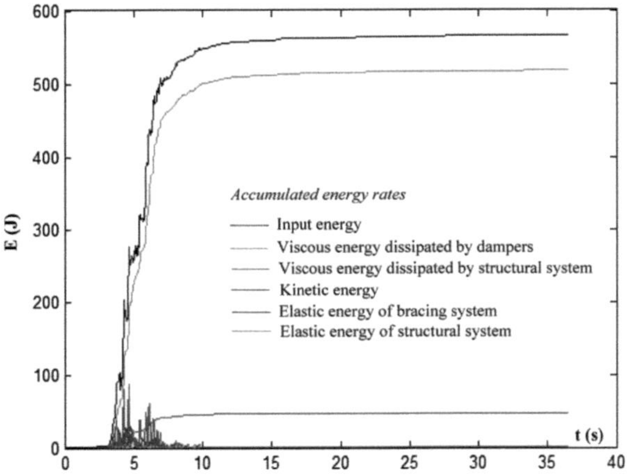

Fig. 6.149 Accumulated energy rates of the MDOF integrated structural system related to the 000055xa-Friuli earthquake

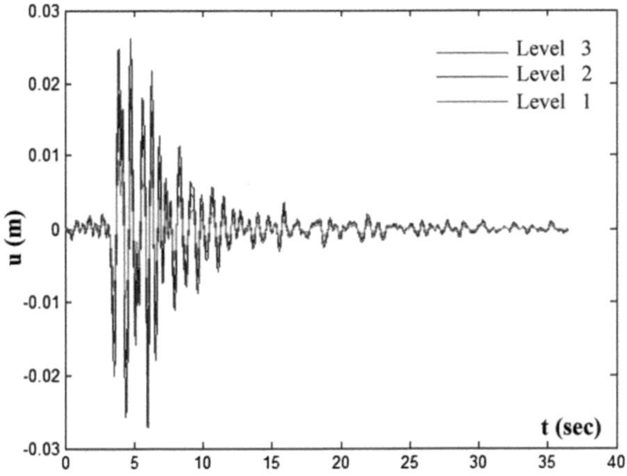

Fig. 6.150 The dynamic response of the MDOF integrated system subjected to the 000055xa-Friuli earthquake

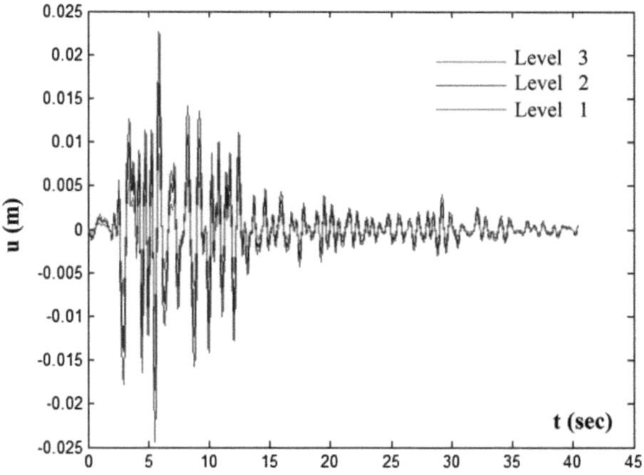

Fig. 6.151 The dynamic response of the MDOF integrated system subjected to the 000198xa-Montenegro earthquake

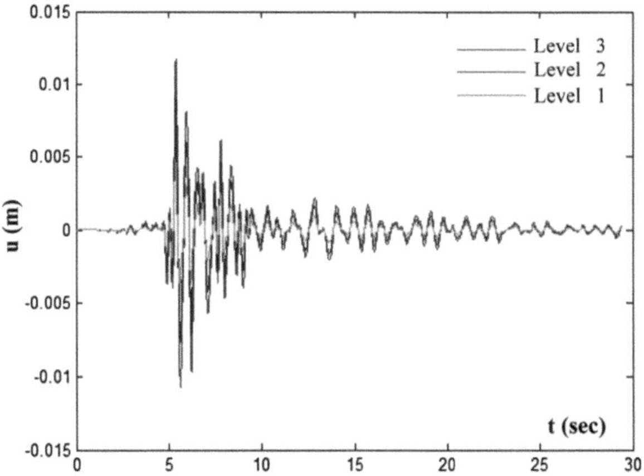

Fig. 6.152 The dynamic response of the MDOF integrated system subjected to the 000665xa-Umbria Marche earthquake

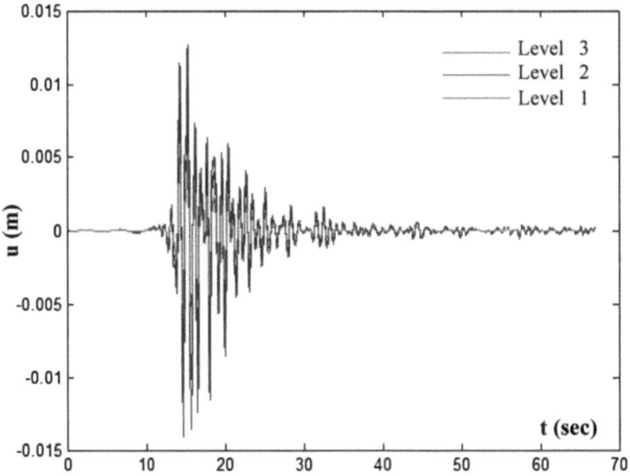

Fig. 6.153 The dynamic response of the MDOF integrated system subjected to the 004675ya-South Iceland earthquake

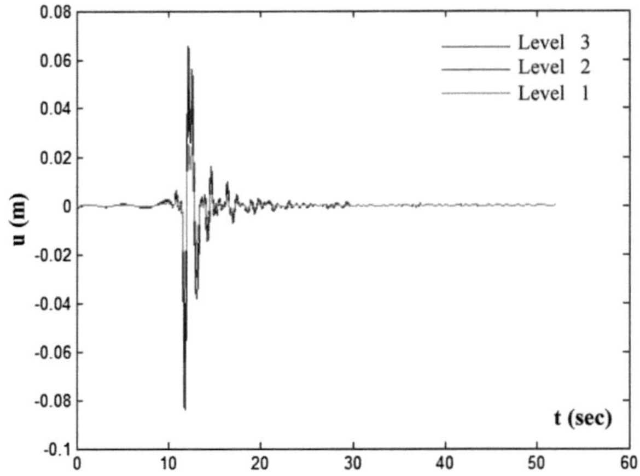

Fig. 6.154 The dynamic response of the MDOF integrated system subjected to the 006332ya-South Iceland (after shock) earthquake

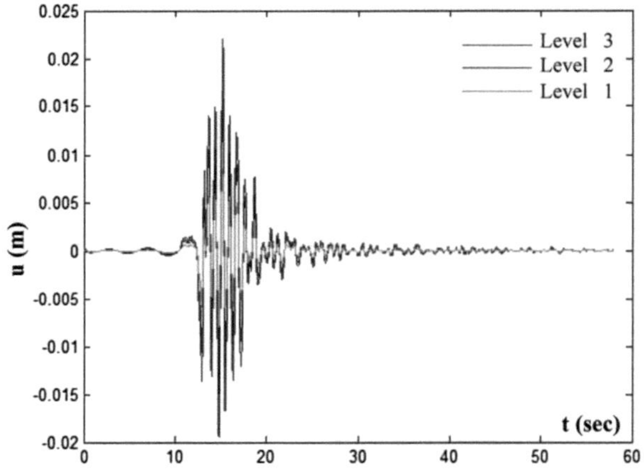

Fig. 6.155 The dynamic response of the MDOF integrated system subjected to the 006335xa-South Iceland (after shock) earthquake

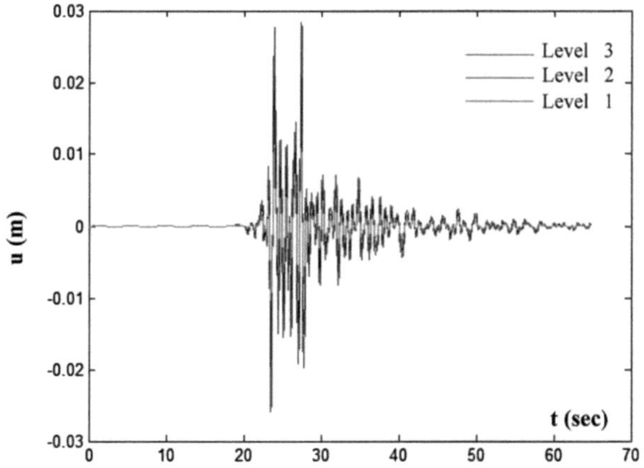

Fig. 6.156 The dynamic response of the MDOF integrated system subjected to the 007142ya-Bingol earthquake

Table 6.4 The dynamic response of the MDOF integrated system subjected to the seven accelerometric recordings selected

Earthquake	u (cm)
000055xa-Friuli	2.71
000198xa-Montenegro	2.43
000665xa-Umbria Marche	1.17
004675ya-South Iceland	1.41
006332ya-South Iceland (after shock)	8.39
006335xa-South Iceland (after shock)	2.21
007142ya-Bingol	2.84
u (average)	3.02

From Figs. 6.142, 6.143, 6.144, 6.145 and 6.147, it is observed that the structure, compared to the previous case, is more deformable at the uppest level and less deformable at the first level. In fact, the system presents, being subjceted to the same force since the mass and the seismic demand are the same, and being characterized by a distribution in elevation of viscosity coefficients proportional to the stiffness values, greater displacements at the uppest level and smaller drifts at the first level. In other words, not being varied seismic demand and global seismic capacity of the integrated structural system, but being varied distribution of its viscoelastic capacity in elevation, the dynamic response of the system in its development in elevation is varied.

With reference to the 000055xa-Friuli accelerometric recording, Fig. 6.148 shows the contributions of the three modes and, as in the previous case, in Fig. 6.149, the accumulated energy rates, evaluated in relative terms, of the input energy, of the kinetic energy of the system, the viscous energy both of the structural system as well as the dissipative bracing system and the elastic energy of both systems are represented for different time values.

Finally, by considering the seven accelerometric recordings, the responses in terms of relative displacements between each level of the MDOF system and the ground are illustrated in Figs. 6.150, 6.151, 6.152, 6.153, 6.154, 6.155 and 6.156.

In Table 6.4, the maximum values of the response of the MDOF integrated structural system subjected to the seven accelerometric recordings considered and the corresponding average value are reported.

Also in this case, the average displacement is equal to 3.02 cm confirming the effectiveness of the integrated design methodology.

6.9 Conclusions

This book, as extensively discussed and illustrated, has explored the possibility of assigning the vibration control, traditionally assigned only to elastic lateral stiffness of the system, even to the viscoelastic resources of a passive dissipation control system.

By following this new integrated design philosophy, the study proposes a simplified method, in accordance to a displacement-based seismic design, aimed at reaching an economically optimal integrated design of elastic structural/vicoselastic control systems which achieves an expected seismic design displacement ranging in a large interval. By assuming that the dynamic behavior of a n-degrees-of-freedom system, characterised by a regularity in elevation, is governed by the first mode, the methodology of the integrated design is developed on an equivalent single-degree-of-freedom structural system. The parametric analysis has been carried out by considering the natural vibration period of the integrated system ranging from 0 to 3 s thus to obtain a spectral representation of the results.

As regards the above-mentioned SDOF system, the design variables considered are, in a first phase, the lateral stiffness of the structural system, the static stiffness and the static viscosity coefficient of the viscoelastic dissipative bracing system by considering a minimum value of the lateral stiffness of the structural system. In a second phase, the natural vibration period of the integrated system has been considered as a design variable and, then, another constraint has also been defined on the maximum value of the overall viscous damping ratio.

The search for the optimal combination of design variables is performed through dynamic analyses on a substitute (SDOF) integrated system by considering a set of seven historical unscaled acceleration records compatible in average with an elastic reference design spectrum.

The optimal design, related to the substitute integrated system corresponding to an expected performance, is represented by the design variables, valued on the average seismic displacement demand related to the acceleration records, that minimize the cost index, assumed as optimized objective function and evaluated for each pair of relative cost ratios considered.

For each of the seven cases analyzed, three design abacuses and their respective contour lines were defined in order to design an optimal integrated system located in an area characterized by the seismic intensity corresponding to the reference displacement average spectrum considered. In fact, from these abacuses, it is possible to obtain the economically optimal values of lateral stiffness of the structural system, the static stiffness and static viscosity coefficient of the visco-elastic dissipative bracing system, for different values of the elastic natural period T and the expected seismic performance *u*.

The methodology consists of defining, for fixed relative cost ratios, the economically optimal combination of the design variables for different values of the period of the single-degree-of-freedom integrated system and the desired performance. In fact, with reference to the relative ratios C_v/C_s equal or lower than 10, for periods longer than 0.5 s and for high performance (small displacements), it is always cheaper to use viscoelastic resources and have a minimum value of lateral stiffness of the structural system necessary for supporting the gravity loads: k_s/k is equal to 0.2. In the case of low performance (high displacements), having only the lateral stiffness of the structural system is economically more convenient. With reference to periods of less than 0.5 s, it is economically advantageous to use the elastic resources of the structural system also in the case of a very high performance.

With reference to the higher relative ratio of the cost C_v/C_s, it can be deduced that the optimal design consists, as it is intuitively expected, of an increasing use of elastic resources of the structural system by reducing the viscoelastic resources of the dissipative bracing system; only for smaller displacements and high periods, is it economically advantageous to use the viscoelastic resources of the dissipative bracing system.

In a second phase, with reference to the relative ratios C_v/C_s equal or lower than 10, relative to the economic optimum of the integrated system obtained by considering the optimal vibration period, it is noted that the cost of the integrated system is always lower than the cost of the other two systems considered showing, thus, the convenience of viscoelastic dissipation in seismic design of structural systems. In such cases, there is a trend for different seismic design displacement values to choose integrated structural systems with a high period and high viscoelastic damping resources resulting the most economically convenient.

In tha case of the higher relative ratio of the cost C_v/C_s, with reference to the economic optimum of the integrated system obtained by considering the optimal vibration period, it is noted that the cost of the integrated system tends to be more and more comparable to that of the structural system with the result that the braced system is the economically optimal solution. In fact, it is observed that there is a regular trend to choose integrated structural systems with more lateral stiffness and with less viscoelastic damping resources for different seismic performance values.

Also in the cases of low relative cost ratios, imposing the upper limit on damping ratio necessarily leads to choose structures with more lateral stiffness (decrease of the economically optimal period) with a consequential increase in costs so that not for all seismic design displacements the integrated design is cheaper than the braced structure. In fact, it is possible to deduce that, unlike the respective analysis without constraint on the damping ratio, there is a more gradual and less marked trend to choose structural integrated systems with a high period and with high viscoelastic damping resources resulting more cost-effective.

Note, however, that these results are relative to the geometric configuration of the viscous damper-brace component with an inclined angle equal to 45°, which is the configuration with the lowest dynamic efficiency. Referring to the other configurations, also for example the one in which the device is horizontal, it follows that the efficiency of the damper is double or more than double and, therefore, the values of the static stiffness would be reduced approximately by 50 % with the consequence that the curves of the costs related to integrated systems would be always lower than the other two curves of the conventional design solutions considered.

An application example of a equivalent single-degree-of-freedom (SDOF) system is used to illustrate the feasibility of the integrated design procedure by using the design abacuses proposed confirming the effectiveness of the integrated design methodology when compared to the other two conventional design solutions.

Finally, the design procedure, developed on a substitute integrated system, has been extended to a proportionally damped MDOF framed integrated system characterised both by a unfirm that variable stiffness distribution in elevation to demonstrate the effectiveness of the integrated design methodology. The equivalence between the SDOF system and the corresponding MDOF one is based on specific hypotheses. In both cases, the integrated system, in accordance to the proposed integrated design methodology, is designed by using the economically optimal values of the stiffness and viscosity coefficient of the structural and viscoelastic dissipative bracing systems for the seismic performance considered. The hypothesis of proportionally damped integrated structural system (i.e., Rayleigh damping) has allowed to evaluate the dynamic response of the integrated system by considering the dynamic behavior of the viscous damper-brace component and taking into account the presence of the stiffness of the brace of each level, or rather, the dependence on the product between the damped frequency of the first mode of the integrated system and the relaxation time correponding to each structural level has been considered to evaluate the dynamic response of each structural level.

Note that by considering a limit on the strength of the vertical elements of the structural system, it should be checked that these elements, designed with that optimal value of lateral stiffness, have adequate strength necessary for supporting the gravity loads. Otherwise, the procedure has to be iterated by increasing the lower limit value of the lateral stiffness of the structural system.

Note that, in the case of the proportionally distributed damping, the linear viscous dampers, unlike viscoelastic, but above all, unlike all the other devices and the two design conventional seismic resistant solutions (unbraced or braced structures), produce forces within a given story that are 90° out of phase with respect to the restoring forces in the same story and, thus, the viscous force is maximum when other forces are zero. This is an exploitable feature especially in the case of existing structures. In fact, the impact of the damping forces on the existing foundation may be minor and therefore the foundations, which are usually very difficult and expensive to retrofit, may require minimal, if any, strengthening.

It is possible to conclude, therefore, that both the fluid viscous and viscoelastic devices are very efficient in reducing the damage to the main structure and often represent economically convenient design solutions.

References

Applied Technology Council (ATC): Seismic evaluation and retrofit of concrete buildings, ATC-40, Redwood City, California (1996)

Applied Technology Council (ATC): NEHRP commentary on the guidelines for the seismic rehabilitation of buildings, Rep. no. FEMA-273, Prepared for the Building Seismic Safety Council (BSSC) by the Applied Technology Council (ATC), Federal Emergency Management Agency (FEMA), Washington D.C. (1997a)

Applied Technology Council (ATC): NEHRP commentary on the guidelines for the seismic rehabilitation of buildings, Rep. no. FEMA-274, Prepared for the Building Seismic Safety Council (BSSC) by the Applied Technology Council (ATC), Federal Emergency Management Agency (FEMA), Washington D.C. (1997b)

Bertero, V.V., Uang, C.-M.: Evaulation of seismic energy in structures. Earthq. Eng. Struct. Dynam. **19**, 77–90 (1992)

Building Seismic Safety Council (BSSC): NEHRP recommended provisions for seismic regulations for new buildings and other structures, 2003 Ed., Rep. Nos. FEMA-450/1 and FEMA-450/2. Prepared by the Building Seismic Safety Council (BSSC) for the Federal Emergency Management Agency (FEMA), Washington, D.C. (2004)

CEN—European Committee for Standardization: Eurocode 8 Part 1: General Rules, Seismic Actions and Rules for Buildings. CEN, Brussels (2005)

Clough R.W., Penzien J.: Dynamics of Structures, 2nd edn. McGraw-Hill, New York (1993)

Fajfar P., Krawinkler H.: Non Linear Seismic Analysis and Design of Reinforced Concrete Buildings. Elsevier, Bled (2005)

Hwang J.S.: Seismic design of structures with viscous dampers. 高科技廠房震害防治研討會, 教育部科技顧問室, September 30, 國立交通大學。, 40–53 (2005)

Hwang J.S., Huang Y.N.: Seismic design of structures with viscous dampers. International training program for seismic design of structures, National Center for Research on Earthquake Engineering, pp. 217–234 (2003)

Hwang, J.-S., Huang, Y.-N., Yi, S.-L., Ho, S.-Y.: Design Formulations for Supplemental Viscous Dampers to Building Structures. J. Struct. Eng. **134**(1), 22–31 (2008)

Miranda, E., Bertero, V.V.: Evaluation of strength reduction factors. Earthq. Spectra Earthq. Eng. Res. Inst. **10**(2), 357–379 (1994)

NTC08: Norme tecniche per le costruzioni, Gazzetta Ufficiale del 04.02.08, DM 14.01.08, Ministero delle Infrastrutture (2008)

Priestley, M.J.N., Seible, F., Calvi, G.M.: Seismic design and retrofit of bridges. Wiley, New York (1996)

Shibata A., Sozen M.A.: Substitute structure method for seismic design in R/C. J. Str. Div. ASCE, **102**(ST1), 1–18 (1976)

Sigaher, A.N., Constantinou, M.C.: Scissor-jack-damper dissipation system. Earthq. Spectra **19** (1), 133–158 (2003)

Wilson E.L.: Three-Dimensional Static and Dynamic Analysis of Structures, 3rd edn. Computers and Structures, Inc., Berkeley (2000)

Index

P. Castaldo, *Integrated Seismic Design of Structure and Control Systems*,
Springer Tracts in Mechanical Engineering, DOI: 10.1007/978-3-319-02615-2,

MIX
Papier aus verantwortungsvollen Quellen
Paper from responsible sources
FSC® C105338

If you have any concerns about our products,
you can contact us on
ProductSafety@springernature.com

In case Publisher is established outside the EU,
the EU authorized representative is:
Springer Nature Customer Service Center GmbH
Europaplatz 3, 69115 Heidelberg, Germany

Printed by Libri Plureos GmbH
in Hamburg, Germany